SpringerBriefs in Applied Sciences and Technology

Computational Intelligence

Series Editor

Janusz Kacprzyk, Systems Research Institute, Polish Academy of Sciences, Warsaw, Poland

SpringerBriefs in Computational Intelligence are a series of slim high-quality publications encompassing the entire spectrum of Computational Intelligence. Featuring compact volumes of 50 to 125 pages (approximately 20,000-45,000 words), Briefs are shorter than a conventional book but longer than a journal article. Thus Briefs serve as timely, concise tools for students, researchers, and professionals.

Fevrier Valdez · Juan Barraza · Patricia Melin

Hybrid Competitive Learning Method Using the Fireworks Algorithm and Artificial Neural Networks

 Springer

Fevrier Valdez
Division of Graduate Studies
Tijuana Institute of Technology, TecNM
Tijuana, Baja California, Mexico

Juan Barraza
Division of Graduate Studies
Tijuana Institute of Technology, TecNM
Tijuana, Baja California, Mexico

Patricia Melin
Division of Graduate Studies
Tijuana Institute of Technology, TecNM
Tijuana, Baja California, Mexico

ISSN 2191-530X   ISSN 2191-5318 (electronic)
SpringerBriefs in Applied Sciences and Technology
ISSN 2625-3704   ISSN 2625-3712 (electronic)
SpringerBriefs in Computational Intelligence
ISBN 978-3-031-47711-9   ISBN 978-3-031-47712-6 (eBook)
https://doi.org/10.1007/978-3-031-47712-6

This Springer imprint is published by the registered company Springer Nature Switzerland AG
The registered company address is: Gewerbestrasse 11, 6330 Cham, Switzerland

Paper in this product is recyclable.

Preface

This book focuses on the fields of artificial neural network, swarm optimization algorithms, clustering problems, and fuzzy logic. This book describes a hybrid method with three different techniques of intelligence computation: artificial neural networks (ANNs), optimization algorithms, and fuzzy logic. Within the artificial neural networks technique, competitive neural network (CNN) is used; for the optimization algorithms technique, we used the fireworks algorithm (FWA), and in the area of fuzzy logic, the Type-1 fuzzy inference systems (T1FIS) and the interval Type-2 fuzzy inference systems (IT2FIS) were used, with their variants of Mamdani and Sugeno type, respectively. It is important to mention that we have proposed new hybrid method to generate new potential solutions in optimization problems in order to find new ways that could improve the results in solving these problems. FWA was adapted for data clustering with the goal to help competitive neural network to find the optimal number of neurons. The metrics used to validate the performance of clusters generated by FWA were intra-cluster and inter-cluster. It is important to mention that two variants were applied to the FWA: dynamically adjust parameters with Type-1 fuzzy logic (FFWA) as the first variant and interval Type-2 (F2FWA) as the second variant. Subsequently, based on the outputs of the CNN and with the goal of classification data, we designed Type-1 and interval Type-2 fuzzy inference systems of Mamdani and Sugeno type, with three variants in the membership functions: triangular, Gaussian, and trapezoidal. The performance of the designed fuzzy classifiers was evaluated with the correct data classification percent. The datasets used to cluster and then to classify were the following: Iris dataset (150 data per 4 features), Wine dataset (178 data per 13 features), and Breast Cancer Wisconsin Diagnostic (569 data per 30 features). It is important to mention that in all case studies, we present the results comparison in order to prove the performance of proposed method.

This book is intended to be a reference for scientists and engineers interested in applying a different metaheuristic or an artificial neural network in order to solve optimization and applied fuzzy logic techniques for solving problems in clustering and classification data. This book can also be used as a reference for graduate courses like the following: soft computing, swarm optimization algorithms, clustering data, fuzzy classification, and similar ones. We consider that this book can also be used to get novel ideas for new lines of research, new techniques of optimization, or to continue the lines of the research proposed by the authors of the book.

In Chap. 1, we begin by offering a brief introduction of the proposed method, the general features of the metaheuristics and its application; also, we present the inspiration, justification, and main contribution of this work.

We describe in Chap. 2 the literature review, basic theoretical and technical concepts about the areas of computational intelligence, and the features of fireworks algorithm and artificial neural network that we need to design the proposed hybrid method. In addition, we presented a brief introduction to the history of swarm optimization algorithms, clustering and classification algorithms, fuzzy logic, and some clustering validation index.

Chapter 3 describes in detail the proposed hybrid method, i.e., we explained in each phases of the development of the hybrid method with competitive learning using FWA and artificial neural network; we also presented equations, examples, explanation, and figures of the proposed hybrid method. It is important to mention that in the clustering validation index, we presented two different ways to implement which are called intra- and inter-cluster validation, respectively.

Chapter 4 is devoted to present the results of the proposed method in different case studies; the first case uses the Iris dataset, the second case uses Wine dataset, and the third case uses Breast Cancer Wisconsin Diagnostic dataset; in the three datasets, we used all features of each data making it more difficult to solve. Finally, we present the design of a fuzzy inference system (FIS), specifically finding the optimal design of a fuzzy classifier. It is important to mention that in all cases studies, we present the results comparison in order to prove the performance of proposed method.

We offer in Chap. 5 the conclusions of this book that are presented at the end in order to mention the advantages of the proposed method and a brief summary of the proposed hybrid method.

We end this preface of the book by thanking all the people who have helped or encouraged us when writing this book. First of all, we would like to thank the respective families for their love and always supporting our work, especially for motivating us to write our research work. We would also like to thank our colleagues working in Soft Computing, which are too many to mention each by their name. Of course, we need to thank our supporting agencies, CONACYT and TNM, in our country for their help during this project. We have to thank our institution, Tijuana Institute of Technology, for always supporting our projects. Finally, we thank our respective families for their continuous support during the time that we spend in this project.

Tijuana, Mexico   Prof. Fevrier Valdez
                                                                              Dr. Juan Barraza
                                                                           Prof. Patricia Melin

Contents

Chapter 1
Introduction to the Hybrid Method Between Fireworks Algorithm and Competitive Neural Network

The Computer Science concept encompasses the combined study of information and computation. We also have learned that the theory of computation describes concepts as algorithms, computation problems, and artificial intelligence, among others [1]. In specific, artificial intelligence [2] is a contrast of the natural intelligence displayed by humans, that is to say, humans have achieved to develop machines that mimic the cognitive Functions based on the human mind, highlighting the most important function which is the learning with the aim that the machines are capable of solving problems [3].

Talking about Artificial Intelligence [4], this discipline has some areas: Optimization Algorithms [5–7], Artificial Neural Network [8], and Fuzzy Logic [9] to mention some.

In the field of Optimization Algorithms, the most famous are the algorithms based on Swarm Optimization[10–13] such as Particle Swarm Optimization (PSO) [13–17], Gravitational Search Algorithm (GSA) [18], Fireworks Algorithm (FWA) [19] etc. The latter we have used to develop a part of this work.

Another field of computer science describes the imitation of biological neural networks of animals or humans and it is denoted and named as ANN and Artificial Neural Networks, respectively.

ANN has two divisions based on their learning. The division of Artificial Neural networks is with supervised and unsupervised learning [20]. Competitive Neural Networks [21] belong to unsupervised learning division and this type of Artificial Neural Network is applied in clustering problems and not in classification problems. However, CNN has a disadvantage, which is determining an arbitrary number of neurons and the disadvantage is considered as a problem; the same problem was attacked with the adaptation of FWA for clustering problems obtained the optimal number of neurons for each dataset used.

On the other hand, and citing another field of computational sciences, the Fuzzy Logic has taken a weighting in the researches of the scientist around the world, that is to say, we have realized the applied of Fuzzy Logic for solving different problems

© The Author(s), under exclusive license to Springer Nature Switzerland AG 2023
F. Valdez et al., *Hybrid Competitive Learning Method Using the Fireworks Algorithm and Artificial Neural Networks*, SpringerBriefs in Computational Intelligence,
https://doi.org/10.1007/978-3-031-47712-6_1

types. Some problems are the following: adjust dynamically of parameters into the algorithm of optimization, control problems, and classification problems to mention some.

After an exhaustive analysis, we combined the three fields of computer sciences mentioned in the previous paragraphs. Therefore, we achieved to design a hybrid method among Competitive Neural Networks (CNNs), Fireworks Algorithm (FWA), and Fuzzy Logic, both Type-1 Fuzzy Logic (T1FL) and Interval Type-2 Fuzzy Logic (IT2FL) [22]. The hybrid method developed is based on expertise, experience, judgement, and other skills of the participating researchers. The method could be partitioned in three phases: in the first phase, FWA is responsible to found the optimal number of neurons for CNN and it can to generate the clustering as the second phase; after, in the third phase, based on mathematics models, the method is capable to design the Type-1 Fuzzy Inference Systems [23, 24] and Interval Type-2 Fuzzy Inference Systems [25] to handle the classification problem.

This book reports the results obtained with three different datasets, Iris, Wine and WDBC datasets. The simulations results are presented in two sections, that is to say, results for the clustering method as the first section and as the second section, for classification problems. As we mentioned above, FWA was adapted to obtain the results in the first section. FWA is converted in the Clustering Fireworks Algorithm (C-FWA) with adaptation in their phases of determining search space using Sturges Law and Root Square of N; and the phase of quality evaluation of their solutions using two different metrics: Inter-cluster and Intra-cluster. For the results of the second section, the method designs Mamdani and Sugeno Types of Fuzzy Inference Systems with Type-1 Fuzzy Logic, and the Interval Type-2 Fuzzy Logic, these systems have three variations in the Membership Functions (MFs) that will be described in the following Chapters.

The structure of this book is described as follows:

Chapter 2 denoted as State of art describes each of the theoretical concepts and shows each of the mathematical and techniques models around the computer sciences to help us the Hybrid method with competitive learning developed.

In Chap. 3, we start by describing the number and features contained in each dataset used in this book. After, we describe a detail each of the phases of the development of the Hybrid method with competitive learning using of FWA Algorithm. In the first phase, we show how the adaptation of FWA for the clustering method was, that is to say, we develop the structure of each spark and fireworks as solutions, and we added on some metrics for defining the bounds and evaluate the quality of solutions. In the second phase shows the participation of the Competitive Neural Network (CNN) to create the clusters and their belonging centroids. In the last phase, we also describe the mathematic models to design the Fuzzy Inference Systems applied in classification problems.

In Chap. 4, the results obtained for all study cases from the proposed hybrid method are presented. Results for clustering and classification problems applied in the Iris, Wine and WDBC datasets, and we also present the statistical comparisons among some methods.

Chapter 5 describes the conclusions based on the results obtained at the end of this book, and also, describes the possible future works with this Hybrid method.

References

1. Wolpert, D.H., Macready, W.G.: No free lunch theorems for optimization. IEEE Trans. Evol. Comput. **1**, 67–82 (1997)
2. Russell, S., Norvig, P.: Artificial Intelligence: A Modern Approach. Prentice-Hall, NJ (2003)
3. Deb, K.: A Population-Based Algorithm-Generator for Real-Parameter Optimization. Springer, Heidelberg (2005)
4. Holland, J.: Adaptation in Natural and Artificial Systems. University of Michigan Press (1975)
5. Antoniou, A., Sheng, W.: In: Antoniou, A., Sheng, W. (eds.) Practical Optimization Algorithms and Engineering Applications: Introduction Optimization, pp. 1–4. Springer (2007)
6. Barraza, J., Melin, P., Valdez, F., González, C.I.: Fireworks algorithm (FWA) with adaptation of parameters using interval type-2 fuzzy logic system. In: Intuitionistic and Type-2 Fuzzy Logic Enhancements in Neural and Optimization Algorithms, pp. 35–47
7. Bonissone, P.P., Subbu, R., Eklund, N., Kiehl, T.R.: Evolutionary algorithms + domain knowledge = real-world evolutionary computation. IEEE Trans. Evol. Comput.Evol. Comput. **10**(3), 256–280 (2006)
8. Hagan, M.T., Demuth, H.B., Beale, M.H.: Neural Network Design. PWS Publishing, Boston, MA (1996)
9. Zadeh, L.A.: Fuzzy logic = computing with words. IEEE Trans. Fuzzy Syst. **4**(2), 103–111 (1996)
10. Eberhart, R., Shi, Y., & Kennedy, J.: Swarm Intelligence. San Mateo, California, Morgan Kaufmann (2001)
11. Engelbrech, P.: Fundamentals of Computational of Swarm Intelligence: Basic Particle Swarm Optimization, pp. 93–129. Wiley (2005)
12. Escalante, H.J., Montes, M., Sucar, L.E.: Particle swarm model selection. J. Mach. Learn. Res. **10**, 405–440 (2009)
13. Shi, Y., & Eberhart, R.C.: A modified particle swarm optimizer. In: Proceedings of the IEEE Congress of Evolutionary Computation, pp. 69–73 (1998)
14. Bonabeau, E., Dorigo, M., Theraulaz, G.: Swarm Intelligence: From Natural to Artificial Systems. OUP USA (1999).
15. Carlisle, A., Dozier, G.: An off-the-shelf PSO. In: Proceedings of the Workshop on Particle Swarm Optimization, pp. 1–6. Indianapolis, USA (2001)
16. Eberhart, R., Shi, Y.: Comparison between genetic algorithms and particle swarm optimization. In: Proceedings of the Seventh Annual Conference on Evolutionary Programming, pp. 611–616 (1998)
17. Kennedy, J., Eberhart, R.: Particle swam optimization. Proc. IEEE Int. Conf. Neural Netw. **4**, 1942–1948 (1995)
18. Sanchez, M.A., Castillo, O., Castro, J.R., Melin, P.: Fuzzy granular gravitational clustering algorithm for multivariate data. Inf. Sci. **279**, 498–511 (2014)
19. Tan, Y.: Fireworks algorithm (FWA). In: Fireworks Algorithm, pp. 17–35. Springer, Berlin Heidelberg (2015)
20. Krogh, A., Vedelsby, J.: Neural network ensembles, cross validation, and active learning. In: Tesauro, G., Touretzky, D., Leen, T. (eds.) Advances in Neural Information Processing Systems, vol. 7, pp. 231–238. MIT Press, Cambridge, MA (1995)
21. Men, H., Liu, H., Wang, L., Pan, Y.: An optimizing method of competitive neural network. Key Eng. Mater. **467–469**, 894–899 (2011)
22. Yao, X., Liu, F.: Evolving neural network ensembles by minimization of mutual information. Int. J. Hybrid Intell. Syst. **1**, 12–21 (2004)

23. Barraza, J., Melin, P., Valdez, F.: Fuzzy FWA with Dynamic Adaptation of Parameters, pp. 4053–4060. IEEE CEC 2016, Vancouver, Canada
24. Zadeh, L.A.: Fuzzy logic. Computer **1**(4), 83–93 (1988)
25. Karnik, N.N., Mendel, J.M.: An Introduction to Type-2 Fuzzy Logic Systems. University of Southern California, Los Angeles, CA (1998)

Chapter 2
Basic Concepts of Neural Networks and Fuzzy Logic

In this chapter, we describe the basic concepts and techniques of computer science, which are crucial to the development of this book.

2.1 Artificial Neural Networks

Neural networks are composed of many elements (Artificial Neurons), grouped into layers that are highly interconnected (with the synapses), which are trained to react (or give values) in a way you want to input stimuli. These systems emulate in some way, the human brain. Neural networks are required to learn to behave (Learning) and someone should be responsible for the teaching or training (Training), based on prior knowledge of the environment problem [1, 2].

A neural network is a system of parallel processors connected together as a directed graph. Schematically, each processing element (neuron) of the network is represented as a node. These connections provide a hierarchical structure trying to emulate the physiology of the brain for processing new models to solve specific problems in the real world. What is important in developing neural networks is their useful behavior by learning to recognize and apply relationships between objects and patterns of objects specific to the real world. In this respect neural networks are tools that can be used to solve difficult problems [3, 4]. Artificial neural networks are inspired by the architecture of the biological nervous system, which consists of a large number of relatively simple neurons that work in parallel to facilitate rapid decision-making [5].

General or basic ways, the Artificial Neural Networks (ANN) are divided depending on your learning type; that is to say, the ANN is divided in two parts: ANN with supervised learning and ANN with unsupervised learning, regardless of the problem to be applied. In the following section, we inquire into about in ANN with unsupervised learning [6, 7].

© The Author(s), under exclusive license to Springer Nature Switzerland AG 2023   5
F. Valdez et al., *Hybrid Competitive Learning Method Using the Fireworks Algorithm and Artificial Neural Networks*, SpringerBriefs in Computational Intelligence,
https://doi.org/10.1007/978-3-031-47712-6_2

2.2 Competitive Artificial Neural Network (CNN)

The artificial neural networks with unsupervised learning have been applied with success in problems of pattern recognition and signal detections. These networks build classes or categories based on input data using correlations or similitude measures and try to identify optimal partitions into the input dataset.

In a competitive neural network, the output units compete with each other to activate; only the one with the highest synaptic potential is activated. The main idea of competitive learning as embodied in the firsts works by von der Malsburg on the self-organization of nerve cells in the cerebral cortex.

Rumelhart and Zisper (1985) specified three basic elements from a rule of competitive learning:

- Neurons set (process units) is activated or not in answer to the input patterns set and they differ in the values of synaptic weights set specific in each neuron.
- A limit imposed on the "strength" of each neuron.
- A mechanism that allows competing among the neurons to respond to the input subset in such a way that only one neuron per the group is activated.

In a simple competitive neural network, individual neurons learn to specialize in sets of similar patterns, and thus become detectors of input pattern characteristics.

In a competitive neural network, a binary process unit is a simple calculation device that only can to present two states: active (on) or inactive (off). The state it presents depends on the signals that come from the input sensors or other process units. Each binary process unit, i, will have associated a synaptic weights vector $w_{i1}, w_{i2}, \ldots w_{iN}$ with it to weighted the values that arrive of the input vectors.

The synaptic potential is defined the following way: If N signals reach the processing unit i, given of the vector $x_1, x_2, \ldots, x_N$ and the synaptic weights vector of each unit is $(w_{i1}, w_{i2}, \ldots, w_{iN})$, then the synaptic potential is given with Eq. 2.1.

$$h_i = w_{i1}x_1 + w_{i1}x_2 + \cdots + w_{iN}x_N - \theta_i \tag{2.1}$$

where:

$$\theta_i = \frac{1}{2}(w_{i1}^2 + w_{i2}^2 + \cdots + w_{iN}^2) \tag{2.2}$$

As a definition, a competitive neural network is established by N input sensors, M process units and connections between each sensor and process unit, so that the connection between sensor j and process unit i have associated a value w_{ij}. Figure 2.1 shows a competitive neural network.

Therefore, if we are representing the status of the process unit i by the variable y_i, it takes the value equal to 1 when it is active and 0 on the contrary, the dynamics of network computing is determined by Eq. 2.3.

Fig. 2.1 CNN architecture

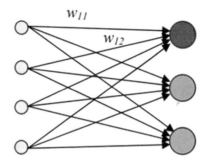

$$y_i = \begin{cases} 1 \; if \quad h_i = \max_{k}\{h_1, h_2, \ldots, h_M\} \\ 0 \; otherwise \end{cases} \quad i = 1, 2, 3, \ldots, M \quad (2.3)$$

The synaptic weights are determined by means of a process of unsupervised learning. It is intended that the process unit whose synaptic weight vector is the most similar to the input vector is activated. The most similarity between the input vector $X = (x_1, x_2, \ldots, x_N)$ and the synaptic weights vector of the process unit i, $W = (w_{i1}, w_{i2}, \ldots, w_{iN})$ it is given by the Euclidian distance between these vectors. The distance is shown in Eq. 2.4.

$$d(x, w_i) = x - w_i = \sqrt{(x_1 - w_{i1})^2 + \cdots + (x_N - w_N)^2} \quad (2.4)$$

Therefore, if r is the winner process unit when the input pattern is introduced, then, the demonstration that the winning unit is the one with the synaptic weight vector most similar to the input vector, is given by Eq. 2.5.

$$d(x, w_r) \leq d(x, w_k), \quad k = 1, 2, \ldots, M \quad (2.5)$$

To continue, we remember that the objective of the competitive neural network with unsupervised learning is to find the groups or classes by itself. For it, the minimum squares criterion is chosen; this criterion is shown in Eq. 2.6.

$$E(k) = \sum_{(i=1)}^{M} a_i k \|x(k) - w_i(k)\|^2 \quad (2.6)$$

where E represent the error in iteration k obtained by the distance between the input pattern $x(k)$ belonging to class M and the synaptic weight vector $x_i(k)$; named as learning rule.

Equation 2.7 shows a detail of the learning rule:

$$w_r(k + 1) = w_r(k) + (k)[x(k) - w_r(k)] \quad (2.7)$$

where the new synaptic weights vector $w_r(k + 1)$ is a combination lineal between the vectors $w_r(k)$ and $x(k)$, and $\eta(k)$ represents the learning rate shown in Eq. 2.8.

$$\eta(k) = {}^{\circ}\eta_0 \left(1 - \frac{k}{T} \right) \tag{2.8}$$

The learning rate will be decrementing throughout the iterations until the value is equal to 0, where k represents the current iteration and T is the total number of the iterations; the initial value of η_0 could be in (0, 1).

As we mentioned in the previous paragraph, the value of η will reaches 0, when these occur, the competitive neural network will stop learning [8].

2.3 Optimization Algorithms

Speaking about optimization, this means to minimize or maximize the solution to a problem depending on a corresponding objective function. We also know that there are different mathematical problems and optimization types: single objective and multi-objective, to mention the most cited by ones researchers [9–12].

At the end of the 80's and the beginning of the 90's the concept of swarm intelligence [13, 14] was created, which was introduced by Gerardo Beni and Jing Wang; nowadays, it is a concept mentioned a lot in the artificial intelligence area, and, swarm intelligence is considered as a branch of artificial intelligence. Swarm intelligence studies the collective behavior of decentralized, self-organized natural or artificial systems [15, 16]. Based on these studies, many optimization algorithms with swarm intelligence have been proposed in recent years, and have been applied to different optimization problems, some examples are: Ant Colony Optimization (ACO) and Particle Swarm Optimization (PSO), which were the first algorithms to come out, then Firefly Algorithm (FA), Fireworks Algorithm (FWA), Grey Wolf Optimizer (GWO) [17, 18] that have come out more recently, just to mention some algorithms of swarm intelligence [16, 19, 20–23].

Thus, as we have been mentioning, for this book, we have used the Fireworks Algorithm (FWA).

2.4 Fireworks Algorithm (FWA)

Another popular algorithm is the particle swarm optimization (PSO) algorithm, which is a population-based search algorithm based on the simulation of the social behavior of birds within a flock. The initial goal of the particle swarm concept was to graphically simulate the graceful and unpredictable choreography of a bird flock [24], this with the aim of discovering patterns that govern the ability of birds to fly synchronously and to suddenly change direction with regrouping in an optimal formation. From this initial objective, the concept evolved into a simple and efficient optimization algorithm.

The Fireworks Algorithm (FWA) is a metaheuristic [3, 25–28] method based on firework explosions, which generate sparks and fireworks representing solutions in the search space. In this Section, the main equations of the algorithm are presented.

To generate sparks as solutions in a search space, Eq. 2.9 is used.

$$Minimize f(x_i) \in R, x_i\_min \le x_i \le x_i\_max \tag{2.9}$$

The locations in the search space are indicated by $x_i = x_1, x_2, x_3, \ldots, x_d$, and the bounds of the same space are represented for $x_{imin} \le x_i \le x_{imax}$ and $f(x_i)$ refers to the objective function [27].

After, the sparks are generated with Eq. 2.10.

$$S_i = m \cdot \frac{y_{max} - f(x_i) + \epsilon}{\sum_{i=1}^{n}(y_{max)} - f(x_i)) + \epsilon} \tag{2.10}$$

For controlling the sparks number, a constant parameter called sparks coefficient, which is denoted as m is used. The worst value (maximum) of the objective function in the n fireworks is represented by $y_{max} = max(f(x_i))(i = 1, 2, 3, \ldots, n)$, and ϵ indicates a constant representing the smallest number, and it is utilized with the goal of avoiding the error in the division by zero [28].

Equation 2.11 defines the bounds that help to maintain the balance in the range of the sparks number. The bounds are defined for s_i.

$$\widehat{S_i} = \begin{cases} round(a \cdot m) & if \quad S_i < am \\ round(b \cdot m) & if \quad S_i > bm, \ a < b < 1, \\ round(S_i) & otherwise \end{cases} \tag{2.11}$$

where a and b are constant parameters [27].

The explosion amplitude for each firework is calculated with Eq. 2.12, but it is important to mention that the explosion is better if the amplitude is small.

$$A_i = \hat{A} \cdot \frac{f(x_i) - y_{min} + \epsilon}{\sum_{i=1}^{n}(f(x_i) - y_{min}) + \epsilon} \tag{2.12}$$

The constant parameter $\widehat{A}$ allows to control the maximum value of the explosion amplitude for each firework, and the best (minimum) value of the objective function among n fireworks is indicated by y_{min} $min(f(x_i))(i = 1, 2, 3, \ldots, n)$, and ϵ, as we mentioned in the description of Eq. 2.10, represents the smallest constant in the computer, and it is used with the goal of not allowing an error of division by zero [25].

Equation 2.13 is using to obtain the location of the sparks.

$$z = round(d \cdot rand(0, 1))$$ (2.13)

The value of d is representing the dimension of the location for each spark and $rand(0, 1)$ is a random value between 0 and 1.

In this algorithm the Eqs. 2.14 and 2.15 are used to select the best current location.

$$R(x_i) = \sum_{j \in K} d(x_i, x_j) = \sum_{j \in K} \| x_i - x_j \|$$ (2.14)

In Eq. 2.14, K is representing the set of all current locations from both fireworks. Equation 2.15 shows how to calculate the probability for selection of a location at x_i.

$$p(x_i) = \frac{R(x_i)}{\sum_{j \in K} R(x_j)}$$ (2.15)

The Angle-based distance, Euclidean distance or Manhattan distance can be used to calculate the distance in the previous equations, which are explained in [27].

2.5　Clustering Algorithms

Following the computational algorithms line, we have seen that there exist a lot of clustering algorithms [29–31], which have been applied to solve classification problems, such as pattern recognition, image segmentation, among other research areas with good results. The famous clustering algorithms are K-Means Clustering [29] and Fuzzy C-Means [30] in classification areas; we can found algorithms as Stochastic Gradient Descent, K-Nearest Neighbors, among others [32, 33].

Also, there are clustering algorithms, which are algorithms with the aim of clustering data based on a specific number of clusters, i.e., if we know the optimal number of clusters for the data, the clustering algorithm would work correctly. However, if we do not know the number optimal of clusters for the data set, the algorithm would not have a good performance, and this feature can be viewed as disadvantage as we mentioned in the introduction [31].

Clustering index validation is very important in the execution of clustering algorithms, thus, these different index clusters validation is described in the following section.

2.6 Clustering Index Validation

In descriptive statistics [34] when data clustering is needed, normally Laws for approximating the optimal number of classes depending on the number of data (N) are used. One of these Laws is called the Sturges Law [35], and it is shown in Eq. 2.16.

$$K = 1 + 3.322 \log N \qquad (2.16)$$

Another Law to estimate the optimal number of clusters is the square root of N, which is indicated in Eq. 2.17.

$$K = \sqrt{N} \qquad (2.17)$$

where K, is the optimal number of clusters and is approximated by using the sample size N in both Eqs. 2.16 and 2.17.

In clustering data, there are two general ways for clustering validation: intra-cluster and inter-cluster. Intra-cluster is the value obtained, by adding the distances of all points with respect to their centroid, and the common distances for Intra-cluster are based on Radius, Radius (variation), Diameter or Diameter (variation) [36] and are represented as shown in Eq. 2.18.

$$Intra = \sum_{i=1}^{n} distance(c_i, C) \qquad (2.18)$$

where c_i are the data that belong to centroid C. Inter-cluster is the value obtained, adding up the distances among the centroids, the common Inter-cluster distance is based on single linkage, complete linkage, UPGM (average distance) or average linkage. In Eq. 2.19, we present a general form to calculate the inter-cluster distance.

$$Inter = \sum_{\substack{i,j=1, \\ i \neq j}}^{k} distance(C_i, C_j) \qquad (2.19)$$

where i and j are the numbers of the centroids, C_i and C_j are different centroids and k is the maximum number of centroids. When we know the optimal number of clusters, it is normal that the best result is with the minimum and maximum distances for Intra-cluster and Inter-cluster, respectively. It is important to mention that the calculation of the distance could be done with Manhattan, Euclidean, Minkowski, Mahalanobis distances or any other distance [37].

2.7 Fuzzy Logic

A fuzzy set A in X is characterized by a Membership (characteristic) function $f_A(x)$, which associates with each point in X a real number in the interval [0, 1], with the value of $f_A(x)$ at x representing the "degree of Membership" of x in A. Thus, the closer the value of $f_A(x)$ to unity, the higher the degree of Membership of x in A. When A is a set in the ordinary sense of the term, its Membership function can take on only two values, 0 or 1, with $f_A(x) = 1$ or 0 according to whether x does or does not belong to A. Thus, in this case, $f_A(x)$ reduces to the familiar characteristic function of set A. When there is a need to differentiate between such sets and fuzzy sets, the sets with two-valued characteristic Functions will be referred to as ordinary sets or simply sets [38–41].

We also know that the fuzzy sets have been applied in different problems, for example, in classification problems. For classification problems, the researches around the world have been using Fuzzy Inference Systems to do the classification [42] of the features with a high degree of similarity for different data sets [43–51].

There are two specific of Fuzzy Inference System types, Mamdani, and Sugeno type. These Fuzzy Inference Systems types have yielded good results in classification problems [52, 53]; both, the application in conjunction with competitive neural networks and the results obtained will be explained in the following sections.

2.8 Interval Type-2 Fuzzy Logic

In this book we described the implementation of Type-2 fuzzy sets as the theory mentions below.

Type-2 fuzzy sets are used to model uncertainty and imprecision; originally, they were proposed by Zadeh [54] and they are essentially "fuzzy–fuzzy" sets in which the Membership degrees are Type-1 fuzzy sets (Fig. 2.2).

The structure of a type-2 fuzzy system implements a nonlinear mapping of input to output space. This mapping is achieved through a set of type-2 if–then fuzzy rules, each of which describes the local behavior of the mapping.

The uncertainty is represented by a region called footprint of uncertainty (FOU). When $\mu_{\tilde{A}}(x, u) = 1, \forall u \in l_x \subseteq [0, 1]$ we have an interval type-2 Membership function (Fig. 2.3) [55].

The uniform shading for the FOU represents the entire interval type-2 fuzzy set and it can be described in terms of an upper Membership function $\overline{\mu}_{\tilde{A}}(x)$ and a lower Membership function $\underline{\mu}_{\tilde{A}}(x)$.

A fuzzy logic system (FLS) described using at least one type-2 fuzzy set is called a type-2 FLS. Type-1 FLSs are unable to directly handle rule uncertainties, because they use type-1 fuzzy sets that are certain [56]. On the other hand, type-2 FLSs are very useful in circumstances where it is difficult to determine an exact certainty value, and there are measurement uncertainties [1, 57–64].

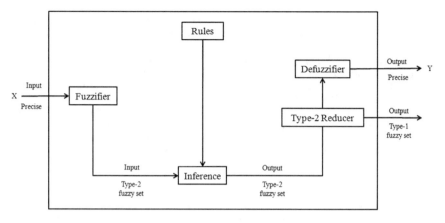

Fig. 2.2 Structure of the interval type-2 fuzzy inference system

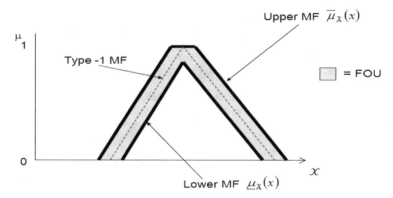

Fig. 2.3 Interval type-2 membership function

References

1. Gaxiola, F., Melin, P., Valdez, F., Castillo, O.: Optimization of type-2 fuzzy weight for neural network using genetic algorithm and particle swarm optimization. In: World Congress on Nature and Biologically Inspired Computing (NaBIC), pp. 22–28 (2013).
2. Zhang, J., Man, K.F.: Time series prediction using recurrent neural network in multi-dimension embedding phase space. IEEE Int. Conf. Syst. Man Cybern. **2**, 11–14 (1998)
3. Tan, Y., Zhu, Y.: Fireworks algorithm for optimization. In: Advances in Swarm Intelligence, vol. 6145, pp. 355–364. Springer-Verlag, Berlin Heidelberg (2010).
4. Yao, X., Liu, F.: Evolving neural network ensembles by minimization of mutual information. Int. J. Hybrid Intell. Syst. **1**, 12–21 (2004)
5. Soltani, S.: On the use of the wavelet decomposition for time series prediction. Neurocomputing **48**(1–4), 267–277 (2002)
6. Akhter, M.R., Arun, A., Sastry, V.N.: Recurrent neural network and a hybrid model for prediction of stock returns. Expert Syst. Appl. **42**, 3234–3241 (2015)

7. Gaxiola, F., Melin, P., Valdez, F., Castillo, O.: Interval type-2 fuzzy weight adjustment for backpropagation neural networks with application in time series prediction. Inf. Sci. **260**, 1–14 (2014)
8. Men, H., Liu, H., Wang, L., Pan, Y.: An optimizing method of competitive neural network optimizing method of com. Key Eng. Mater. **467–469**, 894–899 (2011).
9. Can, U., Alatas, B.: Physics based metaheuristic algorithms for global optimization. Am. J. Inf. Sci. Comput. Eng. **1**, 94–106 (2015)
10. Zheng, Y., Song, Q., Chen, S.-Y.: Multiobjective fireworks optimization for variable-rate fertilization in oil crop production. Appl. Soft Comput. **13**, 4253–4263.
11. Soto, J., Melin, P.: Optimization of the Fuzzy Integrators in Ensembles of ANFIS Model for Time Series Prediction: The Case of Mackey-Glass. IFSA-EUSFLAT, pp. 994–999 (2015).
12. Whitley, L.D.: Foundations of Genetic Algorithms 2. Morgan Kaufman Publishers, pp. 1–332 (1993).
13. Barraza, J., Melin, P., Valdez, F., González, C.I.: Optimal number of clusters finding using the fireworks algorithm. Hybrid Intell. Syst. Control Pattern Recogn. Med. 83–93 (2020).
14. Eberhart, R., Shi, Y., Kennedy, J.: Swarm Intelligence. Morgan Kaufmann, San Mateo, California (2001).
15. Chiou, Y.C., Lan, L.W.: Genetic fuzzy logic controller: an iterative evolution algorithm with new encoding method. Fuzzy Sets Syst. **152**(3), 617–635 (2005)
16. Davis, L.: Handbook of Genetic Algorithms. Van Nostrand Reinhold (1991)
17. Rodriguez, L., Castillo, O., Soria, J.: Grey Wolf Optimizer (GWO) with Dynamic Adaptation of Parameters, pp. 3116–3123. IEEE CEC (2016).
18. Rodriguez, L., Castillo, O., Soria, J.: A study of parameters of the grey wolf optimizer algorithm for dynamic adaptation with fuzzy logic. In: Nature-Inspired Design of Hybrid Intelligent Systems, pp. 371–390 (2017).
19. Aladwan, F., Alshraideh, M., Rasol, M.: A genetic algorithm approach for breaking of simplified data encryption standard. Int. J. Security Appl. **9**(9), 295–304 (2015)
20. Buckles, B.P., Petry, F.E.: Genetic Algorithms. IEEE Computer Society Press (1992).
21. Goldberg, D.E. (1989). Genetic Algorithms in Search, Optimization, and Machine Learning. Addison-Wesley Publishing Company (1989).
22. Parsopoulos, K.E., Vrahatis, M.N.: Particle Swarm Optimization Intelligence: Advances and Applications, pp. 18–40. Information Science Reference, USA (2010)
23. Reeves, C., Row, J.: Genetic Algorithms: Principles and Perspectives, pp. 4–8. Kluwer Academic Publishers, New York, Boston, Dordrecht, London (2002)
24. Clerc, M., Kennedy, J.: The particle swarm-explosion, stability, and convergence in a multimodal complex space. IEEE Trans. Evol. Comput.Evol. Comput. **6**, 58–73 (2002)
25. Abdulmajeed, N.H., Ayob, M.: A firework algorithm for solving capacitated vehicle routing problem. Int. J. Adv. Comput. Technol. **6**(1), 79–86.
26. Melián, B., Moreno, J.: Metaheurísticas: una visión global. Revista Iberoamericana de Inteligencia Artificial **19**, 7–28 (2003)
27. Tan, Y.: Fireworks algorithm (FWA). In: Fireworks Algorithm, pp. 17–35. Springer-Verlag Berlin Heidelberg (2015).
28. Tan, Y., Zheng, S.: Dynamic Search in Fireworks Algorithm, pp. 3222–3229. Evolutionary Computation (CEC 2014), Beijing.
29. Nerurkar, P., Shirkete, A., Chandanec, M., Bhirudd, S.: A novel heuristic for evolutionary clustering. In: Procedia Computer Science 2018, vol. 125, pp. 780–789. Kurukshetra, India.
30. Sanchez, M.A., Castillo, O., Castro, J.R., Melin, P.: Fuzzy granular gravitational clustering algorithm for multivariate data. Inf. Sci. **279**, 498–511 (2014)
31. Soler, J., Tenće, F., Gaubert, L., Buche, C.: Data clustering and similarity. In: Proceedings of the Twenty-Sixth International Florida Artificial Intelligence Research Society Conference, pp. 492–495 (2013).
32. Chena, X., Liua, S., Chena, T., Zhangb, Z., Zhangb, H.: An improved semi-supervised clustering algorithm for multi-density datasets with fewer constraints. Proc. Eng. **29**, 4325–4329 (2012)

33. Rubio, E., Castillo, O.: Interval Type-2 fuzzy possibilistic c-means optimization using particle swarm optimization. In: Nature-Inspired Design of Hybrid Intelligent Systems, pp. 63–78 (2017).
34. Larson, R., Farber, B.: Elementary Statistics Picturing the World, pp. 428–433. Perarson Education Inc. (2003).
35. Sturges, H.A.: The choice of a class interval. J. Am. Stat. Assoc. 21(153), 65–66 (1926).
36. Telescaa, L., Bernardib, M., Rovellib, C.: Intra-cluster and inter-cluster time correlations in lightning sequences. Phys. A 356, 655–661 (2005)
37. Wu, B.Y.: On the intercluster distance of a tree metric. In: Theoretical Computer Science, vol. 369, pp. 136–141. Elsevier Science Publishers Ltd. Essex, UK (2006).
38. Zadeh, L.A.: Fuzzy logic = computing with words. IEEE Trans. Fuzzy Syst. 4(2), 103–111 (1996)
39. Zadeh, L.A.: Fuzzy logic, neural networks and soft computing. Commun. ACM. ACM 37(3), 77–84 (1994)
40. Zadeh, L.A.: Fuzzy logic. Computer 1(4), 83–93 (1988)
41. Zadeh, L.A.: Knowledge representation in fuzzy logic. IEEE Trans. Knowl. Data Eng.Knowl. Data Eng. I, 89–0084 (1989)
42. Maclin, R., Shavlik, J.W.: Combining the predictions of multiple classifiers: using competitive learning to initialize neural networks. In: Proceedings of IJCAI-95, Montreal, Canada, pp. 524–530. Morgan Kaufmann, San Mateo, CA (1995).
43. Ascia, G., Catania, V., Panno, D.: An integrated fuzzy-GA approach for buffer management. IEEE Trans. Fuzzy Syst. 14(4), 528–541 (2006)
44. Horikowa, S., Furuhashi, T., Uchikawa, Y.: On fuzzy modeling using fuzzy neural networks with the backpropagation algorithm. IEEE Trans. Neural Netw. 3, 801–806 (1992).
45. Ishibuchi, H., Nozaki, K., Yamamoto, N., Tanaka, H.: Selecting fuzzy if-then rules for classification problems using genetic algorithms. IEEE Trans. Fuzzy Syst. 3, 260–270 (1995)
46. Jang, J.S.R.: ANFIS: adaptive-network-based fuzzy inference systems. IEEE Trans. Syst. Man Cybern. 23, 665–685 (1992).
47. Jang, J.S.R.: Fuzzy modeling using generalized neural networks and Kalman filter algorithm. In: Proceedings of the Ninth National Conference on Artificial Intelligence (AAAI-91), pp. 762–767 (1991).
48. Mamdani, E.H., Assilian, S.: An experiment in linguistic synthesis with a fuzzy logic controller. Int. J. Man-Mach. Stud. 7, 1–13 (1975)
49. Pedrycz, W.: Fuzzy Evolutionary Computation. Kluwer Academic Publishers, Dordrecht (1997)
50. Pedrycz, W.: Fuzzy Modelling: Paradigms and Practice. Kluwer Academic Press, Dordrecht (1996)
51. Simoes, M., Bose, K., Spiegel, J.: Fuzzy logic based intelligent control of a variable speed cage machine wind generation system. IEEE Trans. Power Electron. 12(1), 87–95 (1997)
52. Takagi, T., Sugeno, M.: Derivation of fuzzy control rules from human operation control actions. In: Proceedings of the IFAC Symposium on Fuzzy Information, Knowledge Representation and Decision Analysis, pp. 55-60 (1983).
53. Takagi, T., Sugeno, M.: Fuzzy identification of systems and its applications to modeling and control. IEEE Trans. Syst. Man Cybern. 15, 116–132 (1985).
54. Karnik, N.N., Mendel, J.M., Qilian, L.: Type-2 fuzzy logic systems. IEEE Trans. Fuzzy Syst. 7(6), 643–658 (1999)
55. Mendel, J.M.: Why We Need Type-2 Fuzzy Logic Systems. In: Mendel, J. (ed.) Article is Provided Courtesy of Prentice Hall (2001).
56. Castillo, O., Melin, P.: Optimization of type-2 fuzzy systems based on bio-inspired methods: a concise review. Inf. Sci. 205, 1–19 (2012)
57. Castro, J.R., Castillo, O., Martínez, L.G.: Interval type-2 fuzzy logic toolbox. Eng. Lett. 15(1), 89–98 (2007)
58. Karnik, N.N., Mendel, J.M.: Applications of type-2 fuzzy logic systems to forecasting of time-series. Inf. Sci. 120, 89–111 (1999)

59. Lee, C.H., Lin, Y.C.: Type-2 fuzzy neuro system via input-to-state-stability approach. In: Liu, D., Fei, S., Hou, Z., Zhang, H., Sun, C. (eds.) LNCS, vol. 4492, pp. 317–327. Springer, Heidelberg (2007).
60. Lee, C.H., Hong, J.L., Lin, Y.C., Lai, W.Y.: Type-2 fuzzy neural network systems and learning. Int. J. Comput. Cogn. **1**(4), 79–90 (2003)
61. Mendel, J.M.: Uncertain rule-based fuzzy logic systems: introduction and new directions. Prentice-Hall, NJ (2001)
62. Wang, C.H., Cheng, C.S., Lee, T.T.: Dynamical optimal training for interval type-2 fuzzy neural network (T2FNN). IEEE Trans. Syst. Man Cybern. Part B Cybern. **34**(3), 1462–1477 (2004).
63. Wu, D., Mendel, J.M.: A vector similarity measure for interval type-2 fuzzy sets and type-1 fuzzy sets. Inf. Sci. **178**, 381–402 (2008)
64. Wu, D., Wan-Tan, W.: Genetic learning and performance evaluation of interval type-2 fuzzy logic controllers. Eng. Appl. Artif. Intell.Intell. **19**(8), 829–841 (2006)

Chapter 3
Hybrid Method Between Fireworks Algorithm and Competitive Neural Network

The main goal of this book is to develop a hybrid method [1–9] of competitive learning using Fireworks Algorithm and a Competitive Neural Network. The proposed method is partitioned into three blocks. We name the first block as Optimization, the second block as clustering, and the third block as Fuzzy classifiers. Figure 3.1 shows the general architecture of the proposed method.

In the first block, the method is capable to find the optimal number of neurons for the competitive neural networks. This optimal number is found with the Fireworks Algorithm (FWA) based on centroids, that is to say, FWA was adapted for the clustering problem and we denoted as FWAC.

After the work done for FWAC, a Competitive Neural Network (CNN) generate centroids as targets, and then, based on some metrics, the method generates clusters based on centroids, in the second block.

In the third block, the method designs the Fuzzy Inference Systems (FIS) as classifiers. The Fuzzy Classifiers were designed with Type-1 Fuzzy Logic and Interval Type-2 Fuzzy Logic, both Mamdani and Sugeno type. The classifiers are based on the clusters generated by the Competitive Neural Network.

In summary, the proposed method sequentially optimizes clusters and classifies a dataset. The three blocks are explained in detail in the following sections.

3.1 Datasets

Firstly, we present the datasets used for this work. The number and features of data with their respective number of classes are presented in Table 3.1.

The characteristics of each dataset are detailed in the following sections [10, 11].

© The Author(s), under exclusive license to Springer Nature Switzerland AG 2023
F. Valdez et al., *Hybrid Competitive Learning Method Using the Fireworks Algorithm and Artificial Neural Networks*, SpringerBriefs in Computational Intelligence,
https://doi.org/10.1007/978-3-031-47712-6_3

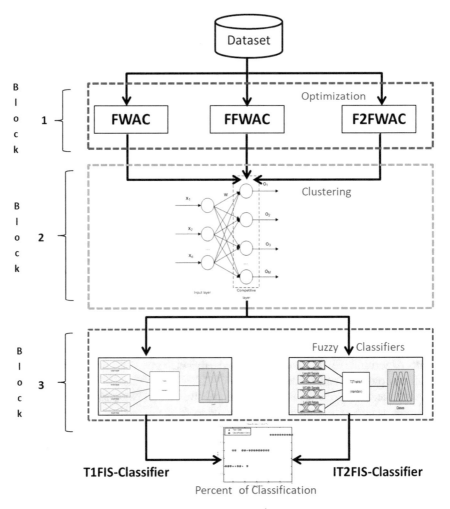

Fig. 3.1 The general proposal architecture

Table 3.1 Datasets structures

Dataset	Data number	Features	Classes
Iris	150	4	3
Wine	178	13	3
WDBC	569	30	2

Table 3.2 Iris dataset features

Number	Features Data		Min value	Max value
1	Length	Sepals	4.3	7.9
2	Width		2	4.4
3	Length	Petals	1	6.9
4	Width		0.1	2.5

3.1.1 Iris Dataset

The Iris data set contains 150 data with four features: Length Sepals, Width Sepals, Length, and Width Petals. The minimum and maximum value per feature are presenting in Table 3.2.

3.1.2 Wine Dataset

The Wine data set contains 178 data with thirteen features. In Table 3.3 the features are presented with their minimum and maximum values, respectively.

Table 3.3 Wine dataset features

Number	Features Data	Min Value	Max Value
1	Alcohol	11.03	14.83
2	Malic acid	0.74	5.8
3	Ash	1.36	3.23
4	Alcalinity of ash	10.6	30
5	Magnesium	70	162
6	Total phenols	0.98	3.88
7	Flavanoids	0.34	5.08
8	Nonflavanoid phenols	0.13	0.66
9	Proanthocyanins	0.41	3.58
10	Color intensity	1.28	13
11	Hue	0.48	1.71
12	OD280/OD315 of diluted wines	1.27	4
13	Proline	278	1680

3.1.3 Breast Cancer Wisconsin Diagnostic (WDBC)

WDBC dataset contains 560 data and thirty features per data, which are presented in Table 3.4.

Table 3.4 WDBC dataset features

Number	Features Data	Min Value	Max Value
1	radius_mean	6.981	28.11
2	texture_mean	9.71	39.28
3	perimeter_mean	43.79	188.5
4	area_mean	143.5	2501
5	smoothness_mean	0.05263	0.1634
6	compactness_mean	0.01938	0.3454
7	concavity_mean	0	0.4268
8	concave_points_mean	0	0.2012
9	symmetry_mean	0.106	0.304
10	fractal_dimension_mean	0.04996	0.09744
11	radius_se	0.1115	2.873
12	texture_se	0.3602	4.885
13	perimeter_se	0.757	21.98
14	area_se	6.802	542.2
15	smoothness_se	0.001713	0.03113
16	compactness_se	0.002252	0.1354
17	concavity_se	0	0.396
18	concave_points_se	0	0.05279
19	symmetry_se	0.007882	0.07895
20	fractal_dimension_se	0.0008948	0.02984
21	radius_largest_worst	7.93	36.04
22	texture_largest_worst	12.02	49.54
23	perimeter_largest_worst	50.41	251.2
24	area_largest_worst	185.2	4254
25	smoothness_largest_worst	0.07117	0.2226
26	compactness_largest_worst	0.02729	1.058
27	concavity_largest_worst	0	1.252
28	concave_points_largest_worst	0	0.291
29	symmetry_largest_worst	0.1565	0.6638
30	fractal_dimension_largest_worst	0.05504	0.2075

3.2 Optimization Methods

In this section, the modification of the Fireworks Algorithm (FWA) for finding the optimal number of clusters is presented, and optimization is the name of the first block in our proposed method.

3.2.1 FWAC

The FWAC [12–14] is a variant of the FWA originally proposed by Tan and Zhu in 2010 [15]. The conventional FWA was proposed to optimize or find the optimal values of some mathematical benchmark Functions, and following in the line of optimization, we implemented this Algorithm (FWA) to automatically find the optimal number of clusters. The reason why we adapted FWA to find the optimal number clusters is because in almost every clustering algorithm, the number of clusters is initialized in a manual or empirical way and is maintained constant while the algorithm is executed. Therefore, it is tedious for researchers to establish the number of clusters for each execution of the algorithm to be validate in each execution and not in each iteration of the algorithm, i.e., depending of the features on data set; we can perform clustering of data with different number of clusters, according to the initial value of the number of clusters.

A problem arises when we do not know the optimal number of clusters, i.e., the perfect situation is when we know the optimal number of clusters for a determined data set, because as is mentioned in some features of clustering algorithms is necessary to know the optimal number of clusters, with the aim that the centroids of the clusters are as best placed as possible, i.e., to consider a good clustering the centroids should be the most separated as possible and in turn, the data should be as compact as possible.

Then, to avoid the inconvenience of not knowing the optimal number of clusters, we decided to adapt the FWA with the main goal of finding the optimal number of clusters as we mentioned above, and this is explained in detail in the operation with the sequence of the following steps: 1. Select n initial locations, 2. Set off n fireworks at n locations, 3. Obtain the locations of sparks, 4. Evaluate the quality of locations and 5. Obtain the new locations.

The main iterations of the Algorithm are between steps 2 and 5, i.e., the algorithm would be iterating while the stopping criterion is not satisfied. In Fig. 3.2 we illustrate the flow chart of the FWAC [12].

To select n initial locations, the first step is generating the initial swarm for the algorithm (FWAC) to start the process; the initial swarm is generated with Eq. 3.1.

$$Swarm_{ij} = LB_j + \left(UB_j - LB_j\right) * r_{ij}, \quad i = 1, 2, 3, \ldots n \qquad (3.1)$$

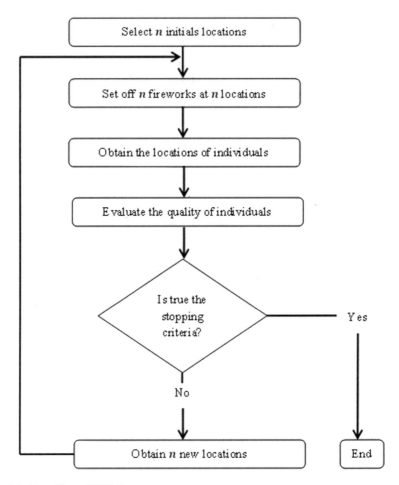

Fig. 3.2 Flow Chart of FWAC

where LB and UB are the lower and upper bounds for individual i in the dimension j, and the dimensions of the individuals depend on the features of the data set given and r is a random value between 0 and 1.

Each individual, in this algorithm, is a spark or a firework; both are represented as a vector divided in two parts, the first part representing the number of centroids and the second part representing the features of the centroids. It is important to mention that the first part consists of integer numbers and second part contains real numbers. In Fig. 3.3, we show an example of a representation of the firework or spark.

To have a better understanding, we are presenting an example of the iris data set, which have 150 data points with four characteristics per data.

In Fig. 3.4, we show an example of how a firework or spark in FWAC is represented, i.e., the vector of possible solutions for this algorithm. In this case K is an integer number that is representing the number of centroids that one possible solution has

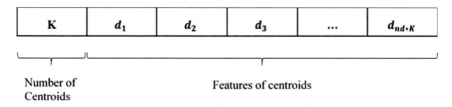

Fig. 3.3 Representation of the individual (firework and spark) in FWAC

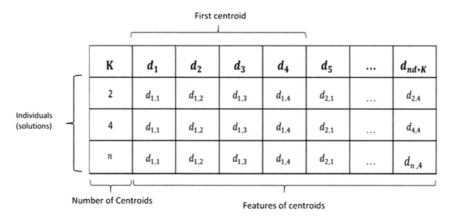

Fig. 3.4 Representation of the individual (firework and spark) in FWAC

and the rest of the vector $(d_1, d_2, d_3, \ldots d_{nd*K})$ are real numbers that are representing the features of the given data set and nd are the features per each datum. Remember that the features per data in the Iris data set are four; however, in the vector, the cells two $(d_{1,1})$, three $(d_{1,2})$, four $(d_{1,3})$, and five $(d_{1,4})$ are representing the features of the first centroid, respectively.

To avoid problems with mathematical calculus and computational processing, after obtaining the vectors to evaluate, we need to convert each vector to matrix form, in specific, the second part of the vector, i.e., the real numbers; these are shown in Fig. 3.5 represented in a general way.

Following the example of the Iris data set and the representation in Fig. 3.5, we present the result of converting vector in matrix solution; the matrix of the first vector is shown in Fig. 3.6 and of the second vector in Fig. 3.7, respectively.

To avoid the overwhelming effect of vector dimensionality, which is based on the number of clusters represented by K, we decided introducing two statistical rules with the goal of creating bounds for the number of clusters, i.e., the rules will allow to obtain a K_{max} value to help the performance of the FWAC. The rules are called Sturges and square root of N; these rules were explained in Chap. 2.

Based on the Iris data set, we show two examples, an example for each statistical rule; we can remember that the numbers of data points is 150 in the mentioned data set.

d_1	d_2	d_3	d_4	...	d_{nd*K}
$d_{1,1}$	$d_{1,2}$	$d_{1,3}$	$d_{1,4}$	...	$d_{1,nd}$
$d_{2,1}$	$d_{2,2}$	$d_{2,3}$	$d_{2,4}$	...	$d_{2,nd}$
$d_{3,1}$	$d_{3,2}$	$d_{3,3}$	$d_{3,4}$	...	$d_{3,nd}$
$d_{4,1}$	$d_{4,2}$	$d_{4,3}$	$d_{4,4}$	...	$d_{4,nd}$
$d_{K,1}$	$d_{K,2}$	$d_{K,3}$	$d_{K,4}$	...	$d_{K,nd}$

Fig. 3.5 General representation of a vector solution converted to a matrix solution in FWAC

K	d_1	d_2	d_3	d_4	d_5	...	d_{nd*K}
2	$d_{1,1}$	$d_{1,2}$	$d_{1,3}$	$d_{1,4}$	$d_{2,1}$	...	$d_{2,4}$
4	$d_{1,1}$	$d_{1,2}$	$d_{1,3}$	$d_{1,4}$	$d_{2,1}$	...	$d_{4,4}$
n	$d_{1,1}$	$d_{1,2}$	$d_{1,3}$	$d_{1,4}$	$d_{2,1}$	...	$d_{n,4}$

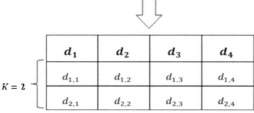

	d_1	d_2	d_3	d_4
$K = 2$	$d_{1,1}$	$d_{1,2}$	$d_{1,3}$	$d_{1,4}$
	$d_{2,1}$	$d_{2,2}$	$d_{2,3}$	$d_{2,4}$

Fig. 3.6 First example of the vector to matrix conversion in FWAC

In Eq. 3.2, we show an example of the Sturges Law with Iris data set.

$$K_{max} = 1 + 3.322 * log\,\mathbf{150} \tag{3.2}$$

We can note that the total result of Eq. 3.2 is equal to 8.2889 (real number), thus, we decided to calculate the floor of the total result to obtain an integer number, in this case, K_{max} which is 8. However, the range of number of clusters for each individual (spark or firework) in FWAC is between 2 and 8, i.e., K_{min} is 2 and K_{max} is 8, and it is important to mention that K_{min} will always be 2 because based on the literature, it is not valid to have just one cluster for the data set.

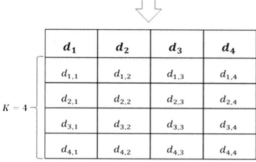

K	d_1	d_2	d_3	d_4	d_5	...	$d_{nd \cdot K}$
2	$d_{1,1}$	$d_{1,2}$	$d_{1,3}$	$d_{1,4}$	$d_{2,1}$	...	$d_{2,4}$
4	$d_{1,1}$	$d_{1,2}$	$d_{1,3}$	$d_{1,4}$	$d_{2,1}$	...	$d_{4,4}$
n	$d_{1,1}$	$d_{1,2}$	$d_{1,3}$	$d_{1,4}$	$d_{2,1}$	...	$d_{n,4}$

	d_1	d_2	d_3	d_4
$K = 4$	$d_{1,1}$	$d_{1,2}$	$d_{1,3}$	$d_{1,4}$
	$d_{2,1}$	$d_{2,2}$	$d_{2,3}$	$d_{2,4}$
	$d_{3,1}$	$d_{3,2}$	$d_{3,3}$	$d_{3,4}$
	$d_{4,1}$	$d_{4,2}$	$d_{4,3}$	$d_{4,4}$

Fig. 3.7 Second example of the vector to matrix conversion in FWAC

The second rule is the square root of N, which is presented in Eq. 3.3.

$$K_{max} = \sqrt{150} \tag{3.3}$$

In the same way as in Eq. 3.2, in Eq. 3.3, we should calculate the floor of the result to obtain an integer number. In particular, for the example of iris data set with the square root of N, K_{max} is 12; accordingly, the range to number of cluster will be between 2 and 12.

Both above mentioned rules, help us approximate the number of clusters based on the total number of data points for each given data set, i.e., based on number of data we can approximate the number of clusters or classes that in FWAC are represented as centroids.

The second step in the FWAC is to finish generating the swarm of solutions, i.e., exploitation of each firework, and in this way, generate the sparks belonging to each firework. As we mentioned in the description of the conventional FWA, the explosion of each firework is based on the explosion amplitude.

To evaluate the swarm of solutions in FWAC is the third step. In previous papers [16, 17], FWA evaluated the swarm of solutions with benchmark mathematical Functions, i.e., each benchmark function has a mathematical expression to be evaluated to obtain a particular fitness. In FWAC, we decide to evaluate with two methods for this problem, which is to find the number of clusters for a data set given.

The two methods that we decided introducing in FWAC to evaluate results are the Intra-cluster and Inter-cluster measures [18]. In Intra-cluster the clustering is validated by calculating the distance between the centroid and its data belonging. On the contrary, in Inter-cluster, only the distance between all centroids is calculated.

In Eq. 3.4 we show how to calculate the fitness when we use the Inter-cluster as a validation measure [19]. As we mentioned before, the distance is calculated between centroids, in this case, C_i and C_j.

$$Fitness = \frac{\sum_{i=1}^{K} \| C_i - C_j \|}{K}, \quad i \neq j. \tag{3.4}$$

First we calculate the distance between all K centroids, after the value is obtained then is divided by the number of centroids to obtain a final value called fitness.

When the fitness is calculated with Intra-cluster [20], the calculation is based on Eq. 3.5.

$$Fitness = \frac{\sum_{i=1}^{K} \sum_{j=1}^{p} a_{ij} \| x(j) - C_i \|}{K} \tag{3.5}$$

where

$$a_{ij} = \begin{cases} 1 \ if \ x(j) \in C_i \\ 0 \ otherwise \end{cases} \tag{3.6}$$

And $x(j)$ are the data that belonging at centroid C_i.

It is important to mention that the problem we are considering is to find the optimal number of clusters based on the centroids for a particular data set, but the task of FWAC is not to find the location of centroids; in this work, we are just searching for the number of clusters.

We explained the main task of FWAC because the optimization of this problem depends on the distances and not the location, thus, we decided to implement the maximum and minimum distance for inter and intra-cluster, respectively. The reason to implement the minimum and maximum distance is because, it is normal to have two centroids with less distance than six centroids with greater distance, but this does not mean that the six centroids are more separate than the two centroids, if we are validating with Inter-cluster. In the same way, if we are validating with Intra-cluster, it is normal to have a greater distance with a cluster with fifteen data points for belonging to a centroid than the minimum distance with a cluster with four data to belong to the centroid, but this does not mean that the cluster with four data points is more compact than the cluster with fifteen data points.

The fourth step of FWAC is to select n new locations in an elitist way, i.e., on the contrary of conventional FWA, where the new locations are selected using the probability of Eq. 2.17, in FWAC, it takes the n best locations in each iteration. The previous step applies if the stop criterion is false. It is important to mention that for distance problems is advisable that the stopping criterion is the maximum number of iterations. The FWAC will be iterating between steps 2 and 4 while the above-mentioned stopping criterion is false.

3.2.2 FFWAC

Modification of FWAC called FFWAC is presented in this section. Based on previous works with FWA[12, 13], we decided to implement the Fuzzy Fireworks Algorithm for Clustering that we denoted as FFWAC. The main difference between FFWAC and FWAC is in the amplitude coefficient parameter (Eq. 2.12), which is a constant value in FWAC and in FFWAC has a value between 2 and 40 determined by fuzzy logic121, thus, Eq. 3.7 is representing the explosion amplitude, but with a dynamic amplitude coefficient.

$$FA_i = \widehat{FA}.\frac{f(x_i) - y_{min} + \epsilon}{\sum_{i=1}^{n}(f(x_i) - y_{min}) + \epsilon} \qquad (3.7)$$

where FA_i is the fuzzy amplitude explosion for each firework and $\widehat{FA}$ has a dynamic value that decreases from 40 to 2 based on the number of function evaluations (Iterations).

In Fig. 3.8 we show the graphical representation of the Fuzzy Inference System for FFWAC [1].

The aim is to adapt the amplitude coefficient in a dynamic way in FFWAC and this is for controlling the algorithm when doing exploration and exploitation [22, 23]. In the initial function evaluations (iterations) in the FFWA, the algorithm will be exploring, in the half of the function evaluations (iterations) the Algorithm will be into the equilibrium point between exploring and exploitation, and finally, in the end of function evaluations (iterations) it will be exploiting.

In FFWAC, we proposed three variations for testing the performance of the Algorithm. The three variations were performed in the Membership Functions of the input and output variables. In the first variation, we used Triangular Membership Functions, the second variation of FFWA has Gaussian Membership Functions and the third variation has Trapezoidal Membership Functions, and we denoted them as FFWAC-I, FFWAC-II and FFWAC-III, respectively.

The variation of the amplitude coefficient in FFWAC is controlled with the following three fuzzy rules [24], which are presented in Fig. 3.9.

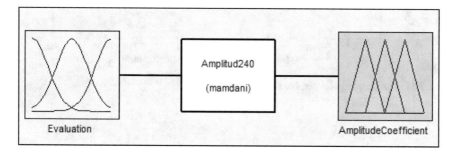

Fig. 3.8 Graphical Representation of the Fuzzy Inference System for FFWAC

1. If (**Evaluation** is **Low**)   then (**AmplitudeCoefficient** is **Big**)
2. If (**Evaluation** is **Medium**) then (**AmplitudeCoefficient** is **Medium**)
3. If (**Evaluation** is **High**)   then (**AmplitudeCoefficient** is **Small**)

Fig. 3.9 Fuzzy Rules for Amplitude Coefficient

The ideas explained above are based on previous works [8, 9] with FWA, i.e., the amplitude coefficient, variations in Membership Functions, fuzzy rules and other concepts. In Figs. 3.10 and 3.11, we show the input and output variables in FFWA-I with Triangular Membership Functions.

As we mentioned before, the Membership Functions of the variables (input and output) in FFWAC-II are Gaussian. The input variable is shown in Fig. 3.12 and the output variable in Fig. 3.13.

Figure 3.14 shows the input variable and Fig. 3.15 shows the output variable with Trapezoidal Membership Functions for FFWAC-III [9, 10].

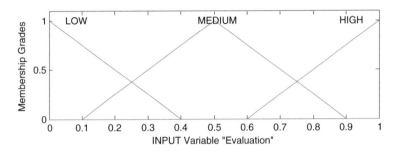

Fig. 3.10 Input triangular variable for FIS of the FFWAC-I

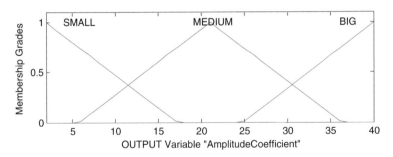

Fig. 3.11 Output triangular variable for FIS of the FFWAC-I

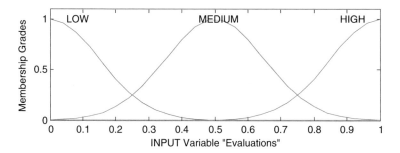

Fig. 3.12 Input Gaussian variable for FIS of the FFWAC-II

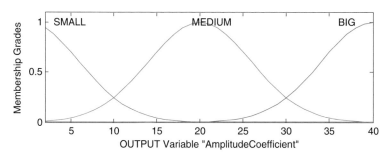

Fig. 3.13 Output Gaussian variable for FIS of the FFWAC-II

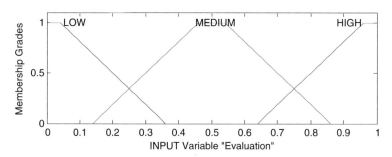

Fig. 3.14 Input Trapezoidal Variable for FIS of the FFWAC-III

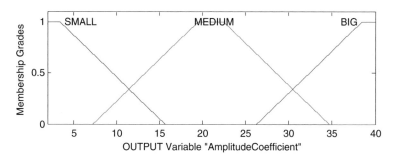

Fig. 3.15 Output Trapezoidal Variable for FIS of the FFWAC-III

3.2.3 F2FWAC

We implemented the Algorithm (FWA) with an adjustment of parameter (Amplitude coefficient) using Interval Type 2 Fuzzy Logic Inference System to automatically find the optimal number of clusters, and we decided to call this as Interval Type 2 Fuzzy Fireworks Algorithm for clustering and we denoted as F2FWAC.

General ways, the steps of the F2FWAC algorithm are the same steps of the FFWAC algorithm but, with the difference that the adjustment of the explosion amplitude parameter in F2FWAC is using Interval Type-2 Fuzzy Logic.

As we mentioned in the above paragraph, we implement an Interval Type 2 Fuzzy Inference System (IT2FIS) for controlling the amplitude coefficient in Eq. 3.8. The modified Equation is the following:

$$FA_i = \widehat{FA}. \frac{f(x_i) - y_{min} + \epsilon}{\sum_{i=1}^{n}(f(x_i) - y_{min}) + \epsilon} \tag{3.8}$$

where $\widehat{FA}$ will be in a range between 2 and 40.

The IT2FIS is shown in Fig. 3.16.

The rules for IT2FIS are the same three rules used in FFWAC (Fig. 3.9).

To test the performance of the F2FWAC we have made three variations in the same, the variations consists in modifying the Membership Functions of the Interval Type-2 Fuzzy Inference Systems (IT2FIS), in concrete, the Membership Functions of the input and output variables [8], below, the three different Membership Functions are explained.

In the first variation of the IT2FIS we used Triangular Membership Functions and for this variation we denoted as F2FWAC-I; in Figs. 3.17 and 3.18, we show the input and output variables, and the ranges for each variable and Membership function in F2FWAC-I.

The linguistic variables, and parameters for each partition in the input variable in F2FWAC-I are the following: Low [− 0.3, − 0.1, 0.1, − 0.1, 0.1 and 0.3], Medium [0.2, 0.4, 0.6, 0.4, 0.6 and 0.8] and High [0.7, 0.9, 1.2, 0.9, 1.1 and 1.4].

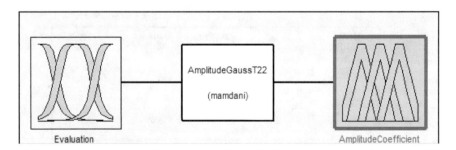

AmplitudeGaussT22

(mamdani)

Evaluation   AmplitudeCoefficient

Fig. 3.16 Graphic representation of the IT2FIS

Fig. 3.17 Input variable in F2FWAC-I

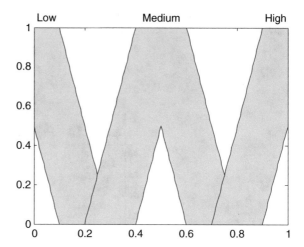

Fig. 3.18 Output variable in F2FWAC-I

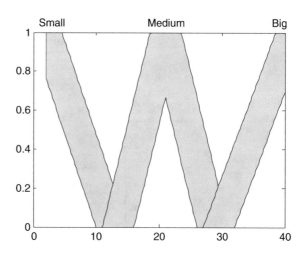

The three partitions of the Output Variable are: Small [− 15, − 0.5, 10, − 10, 4.5 and 15], Medium [11, 18.5, 26, 16, 23.5 and 31] and Big [27, 38.5, 45, 32, 43.5 and 50].

In Figs. 3.19 and 3.20, the second variation of the F2FWAC is presented.

The parameters of the Gaussian Membership Functions for F2FWAC-II in the input are: Low [0.1, − 0.1 and 0.1], Medium [0.1, 0.4 and 0.6] and High [0.1, 0.9 and 1.1] and for the output variable are: Small [5, − 5 and 5], Medium [5, 16 and 26] and Big [5, 36 and 46].

For the third variation we implemented Trapezoidal Membership Functions in the variables of the IT2FIS, and this variation was denoted as F2FWAC-III.

In Figs. 3.21 and 3.22, the parameters of the Membership Functions for the input and output variables are shown.

Fig. 3.19 Input variable for F2FWAC-II

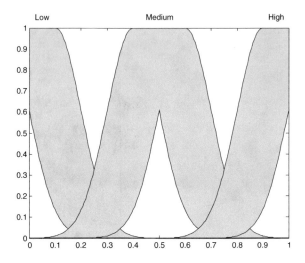

Fig. 3.20 Output variable for F2FWAC-II

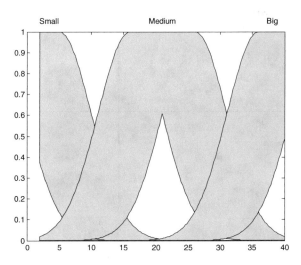

Low [− 0.4, − 0.1, 0.01, 0.3, − 0.3, 0.01, 0.1, 0.4 and 0.9], Medium [0.01, 0.3, 0.5, 0.8, 0.1, 0.4, 0.6, 0.9 and 0.9] and High [0.5, 0.8, 1.01, 1.3, 0.6, 0.9, 1.1, 1.4 and 0.9] are the partitions and parameters for the input variable in F2FWAC-III using Trapezoidal Membership Functions.

The output variable in IT2FWAC-III has the following partitions and ranges: Small [− 11.6, − 3.6, 3.3, 13, − 7.2, − 3.1, 8.8, 19 and 0.9], Medium [2.5, 14, 21, 33, 8, 20, 27, 39 and 0.9] and Big [21, 33, 40, 52, 27, 39, 46, 58 and 0.9].

Fig. 3.21 Input variable for F2FWAC-III

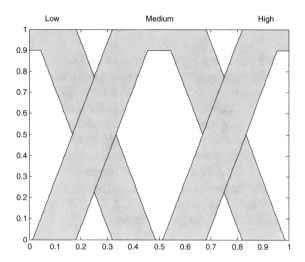

Fig. 3.22 Output variable for F2FWAC-III

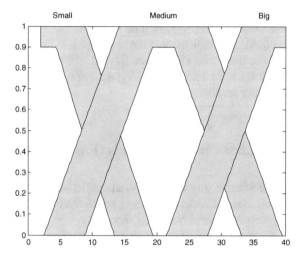

3.3 Clustering Method

The second block is named as Clustering, in this block, a Competitive Neural Network provides us the centroids for each dataset, and we introduced a metric to generate the clusters based on centroids. It is explained in more detail below.

3.3.1 Competitive Neural Network

As we mentioned in the literature review section, a competitive neural network has unsupervised learning, i.e., the neural network should be learning based on the input data; in this case, the conventional competitive neural network does not know the target or targets will be presented in the problem.

In our proposed method, the main goal of the competitive neural network (CNN) will be to create clusters based on the input data, for the consequence, the targets of the neural network will function as centroids of their clusters.

CNN will have a conventional functionally as we explained in previous sections. We take the targets as centroids and below, Eq. 3.9 shows as the cluster is formed.

For each data x_i, where $i = 1, 2, 3, \ldots N$ we are going to calculating the distance to each Centroid (targets) with Eq. 3.9.

$$dist = \|x_i - C_n\| \tag{3.9}$$

where $n = 1, 2, \ldots, M$; N represents the total input data and M the total numbers of neurons in competitive neural network. The minimum distance among de data x_i and Centroids, indicate that data belonging to the centroid C_n.

Once the clusters have been formed with their data belonging, the next step is to design the Type 1 and Interval Type 2 Fuzzy Inference Systems as we explain in the following section.

3.4 Classification Methods (Fuzzy Classifiers)

In this Section, we explain in detail how to automatically design Type-1 and Interval Type-2 Fuzzy Inference Systems, respectively; based on the clusters formed by the Competitive Neural Network, previously.

We start to say that the design of Fuzzy Inference Systems will be creating with all features of the input dataset, that is to say, input variables (Membership Functions), rules and outputs; will depending on the number input data and features number.

The number of input variables in each Fuzzy Inference System will be depending on the number of features that the data contained in each dataset; and the number of Membership Functions for each input variable, will be depending on the number of clusters generated by the competitive neural network.

In this work, we achieve to design Fuzzy Inference Systems with three different Membership Functions: Triangular, Gaussian and Trapezoidal. The details and characteristics of each Fuzzy Inference System designed are presented, below.

3.4.1 T1FIS-Mamdani and Sugeno Type

In the first step, we explained the design of Membership Functions for Type Fuzzy Inference Systems of Mamdani and Sugeno type.

3.4.1.1 Triangular Membership Functions

To generate a Triangular Membership function is necessary to have three points a, b and c, with the restriction of $a < b < c$; as we shown in Eq. 3.10.

$$Triangle(x; a, b, c) = \begin{cases} 0, & x \leq a. \\ \dfrac{x - a}{b - a}, & a \leq x \leq b. \\ \dfrac{c - x}{c - b}, & b \leq x \leq c. \\ 0, & c \leq x. \end{cases} \tag{3.10}$$

But, as we have mentioned, the number of Membership Functions will depend on the number of clusters; and, an input variable will have n Membership Functions based on the dataset used in the problem.

In Fig. 3.23 we are presenting a model of a Triangular Membership function with the three point (a, b and c) mentioned above, for the design of the proposed method. Where for each feature (dimension) of data:

- To obtain the point a_n, we will subtract two standard deviations of the cluster$_n$, to the position of the Centroid$_n$ the same cluster.
- The point b_n, will be the position of the centroid of cluster$_n$.
- To obtain de point c_n, we will add two standard deviations of the cluster$_n$ to the position of the Centroid$_n$ the same cluster.

In Table 3.5, we present in detail how to design a Triangular Membership function. The Triangular Membership function 1, will be formed with the points a_1, b_1 and c_1, respectively. The Triangular Membership function 2, will be formed with the

Fig. 3.23 Design of a Triangular Membership function

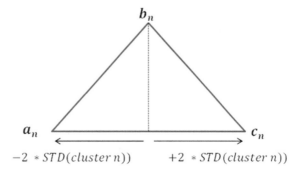

Table 3.5 Points for Triangular Membership Functions

$Memebership Function_1$	$Memebership Function_2$	...	$Memebership Function_n$
$a_1 = Centroide_1 - 2 * STD(Cluster_1)$	$a_2 = Centroide_2 - 2 * STD(Cluster_2)$	$\cdots$	$a_n = Centroide_1 - 2 * STD(Cluster_n)$
$b_1 = Centroide_1$	$b_2 = Centroide_2$	...	$b_n = Centroide_n$
$c_1 = Centroide_1 + 2 * STD(Cluster_1)$	$c_2 = Centroide_2 + 2 * STD(Cluster_2)$	$\cdots$	$c_n = Centroide_1 + 2 * STD(Cluster_n)$

points a_2, b_2 and c_2; and so on, Triangular Membership function n, will be formed with the points a_n, b_n and c_n. Knowing that n represents the total of the numbers of Memberships function for each input variable, which will depend on the problem to be addressed.

3.4.1.2 Gaussian Membership Functions

Normally, in fuzzy sets, to generate a Gaussian Membership function is necessary to have two points c and σ, where c represents the media and σ the standard deviation. Equation 3.11 shows the mathematical model to generate a Gaussian Membership function.

$$Gaussian(x; c, \sigma) = e^{-\frac{1}{2}\left(\frac{x-c}{\sigma}\right)} \tag{3.11}$$

In Fig. 3.24, we show the model of a Gaussian Membership for the proposed method.

Where for each feature (dimension) of data:

- The point C_n, we will represent by position of the $Centroid_n$ of $cluster_n$.
- The point σ_n will be obtained, calculated the standard deviation of the $cluster_n$.

The Gaussian Membership function 1, will be formed with the points c_1 and σ_1, respectively; and so on, Gaussian Membership function n, will be formed with the points c_n and σ_n.

Fig. 3.24 Design of a Gaussian Membership function

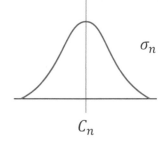

Table 3.6 Points for Gaussian Membership Functions

$Memebership Function_1$	$Memebership Function_2$	...	$Memebership Function_n$
$c_1 = Centroid(Cluster_1)$	$c_2 = Centroid(Cluster_2)$	...	$c_n = Centroid(Cluster_n)$
$\sigma_1 = STD(Cluster_1)$	$\sigma_2 = STD(Cluster_2)$	...	$\sigma_n = STD(Cluster_n)$

STD is an abbreviation of Standard Deviation

In Table 3.6, we explained the two points mentioned above, which points to help us how to design a Gaussian Membership function.

3.4.1.3 Trapezoidal Membership Functions

In fuzzy sets, is necessary to have four points to generate a Trapezoidal Membership function. We list the four points as a, bc and d; as written in Eq. 3.12.

$$Trapezoidal(x; a, b, c, d) = \begin{cases} 0, & x \leq a. \\ \dfrac{x - a}{b - a}, & a \leq x \leq b. \\ 1, & b \leq x \leq c. \\ \dfrac{d - x}{d - c}, & c \leq x \leq d. \\ 0, & d \leq x. \end{cases} \qquad (3.12)$$

With the mathematical restriction that $a < b \leq c < d$.

In the proposed method, we introduced statistics method, a summary of the five numbers (variation measure) to obtain the box and mustache graphic, and thus generate, automatically, the design of a Trapezoidal Membership function as shown in Fig. 3.25.

As we can see in Fig. 3.25, with the graphic we have obtained the four necessary points which are: a_n, b_n, c_n and d_n, respectively.

Table 3.7 describes in detail the obtaining of points mentioned above.

With the summary of five numbers, we are obtaining: minimum value, Quartile #1, Quartile #2, Quartile #3 and maximum value, of which, the minimum value of $Cluster_n$ represents the point a_n, Quartile #1 and #3 of the $Cluster_n$ represents the points b_n and c_n, respectively; and d_n is represented by the maximum value of the $Cluster_n$.

The **fuzzy rules** for the proposed method will also be generated automatically that is to say, the rules are depending on the number of the input data and the features number of each data.

In any Fuzzy Inference System, the rules are depending on the inputs variables and the partition of input variables (number of Membership Functions). The total number of fuzzy rules is calculated multiplying the number of input variables by the number of Membership Functions; it is necessary, the researchers could be using the total number of fuzzy rules or, in a specific case, could be used fewer rules than the

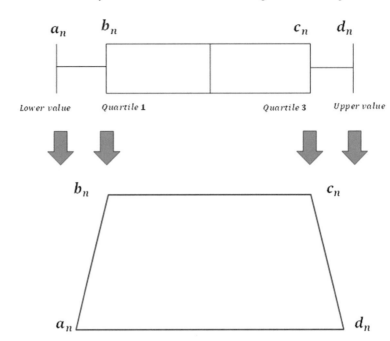

Fig. 3.25 Design of a Trapezoidal Membership function

Table 3.7 Points for trapezoidal membership functions

$Memebership Function_1$	$Memebership Function_2$	...	$Memebership Function_n$
$a_1 = min(Cluster_1)$	$a_2 = min(Cluster_2)$	...	$a_n = min(Cluster_n)$
$b_1 = Quartile1(Cluster_1)$	$b_2 = Quartile1(Cluster_2)$	...	$b_n = Quartile1(Cluster_n)$
$c_1 = Quartile3(Cluster_1)$	$c_2 = Quartile3(Cluster_2)$	...	$c_n = Quartile3(Cluster_n)$
$d_1 = max(Cluster_1)$	$d_2 = max(Cluster_2)$	...	$d_n = max(Cluster_n)$

total possible number of fuzzy rules. The following list, it is explained as to generate the fuzzy rules for the T1FISs:

1. **IF** ($Input_1$ is $Cluster_1$) and ($Input_2$ is $Cluster_1$) and,…, ($Input_n$ is $Cluster_1$) **THEN** ($Output$ is $Class_1$).
2. **IF** ($Input_1$ is $Cluster_2$) and ($Input_2$ is $Cluster_2$) and,…, ($Input_n$ is $Cluster_2$) **THEN** ($Output$ is $Class_2$).
3. **IF** ($Input_1$ is $Cluster_3$) and ($Input_2$ is $Cluster_3$) and,…, ($Input_n$ is $Cluster_3$) **THEN** ($Output$ is $Class_3$).

⋮

m. **IF** ($Input_1$ is $Cluster_k$) and ($Input_2$ is $Cluster_k$) and,…, ($Input_n$ is $Cluster_k$) **THEN** ($Output$ is $Class_k$).

Where m represents the total number of rules, n the number of input data and k, the events number to evaluate, these last one could be, values predictions, classification of classes, etc.

For the proposed method, as we mentioned in the above paragraph, depending on the applied problem, the rules and the consequence, the output variables could vary; for example, if the T1FIS is applying to a classification problem, the output will depend on the number of classes to be classify.

In fact, the design of the Fuzzy Inference Systems of Mamdani and Sugeno type are very similar, the specific difference is contained in the output variables, that is to say, into the FISs of Mamdani type the output variables has Membership Functions partitions, on contrary, into the FISs of Sugeno the output variables have a constant parameter or a mathematical function.

For this book, we have used the classification problem to evaluate the performance of the models; in the following section, we are going to give more details.

3.4.2   IT2FIS-Mamdani and Sugeno Type

As we have known, the Interval Type 2 Fuzzy Inference Systems (IT2FIS) has a more uncertainty degree than the Type 1 Fuzzy Inference Systems. Thus, in the following three sections we explain, how we designing the IT2FIS.

We will start by mentioning the parameter that we introduced in the design of the IT2FIS Membership Functions. The error parameter is shown in Eq. 3.13.

$$error = \frac{\left| w_{ij} - w_{kj} \right|}{2} \tag{3.13}$$

The error is the absolute distance between the initial position and the final position of the centroid belonging from the cluster divided by two. Therefore, w is the centroid, i is initial position, k is the final position and j is the dimension (feature data for this work).

3.4.2.1   Triangular Membership Functions

In Sect. 3.4.1.1 we explained the design of Triangular Membership Functions for T1FIS. Triangular Membership Functions for IT2FIS is designing with six points, which are: $a, b, c, d, e,$ and f; Table 3.8 shows the calculus for each point.

In Fig. 3.26, graphic representation of a Triangular Membership function is shown. Black triangle represents a Triangular Membership function for T1FIS and, red triangles represent the Triangular Membership function for IT2FIS.

Table 3.8 Calculus of the points in Triangular Membership Functions for IT2FIS

Membership function	Formula
	$a_1 = (centroide - 2 * STD(cluster\ 1)) - error$
	$b_1 = centroide(cluster\ 1) - error$
$Membership Function_1$	$c_1 = (centroide + 2 * STD(cluster\ 1)) - error$
	$d_1 = (centroide - 2 * STD(cluster\ 1)) + error$
	$e_1 = centroide(cluster\ 1) + error$
	$f_1 = (centroide + 2 * STD(cluster\ 1)) + error$
	$a_2 = (centrode - 2 * STD(cluster\ 2)) - error$
	$b_2 = centroide(cluster\ 2) - error$
$Membership Function_2$	$c_2 = (centroide + 2 * STD(cluster\ 2)) - error$
	$d_2 = (centroide - 2 * STD(cluster\ 2)) + error$
	$e_2 = centroide(cluster\ 2) + error$
	$f_2 = (centroide + 2 * STD(cluster\ 2)) + error$
…	$a_n = (centroide - 2 * STD(cluster\ n)) - error$
	$b_n = centroide(cluster\ n) - error$
$Membership Function_n$	$c_n = (centroide + 2 * STD(clustern)) - error$
	$d_n = (centroide - 2 * STD(clustern)) + error$
	$e_n = centroide(cluster\ n) + error$
	$f_n = (centroide + 2 * STD(cluster\ n)) + error$

Fig. 3.26 Graphic representation of Triangular Membership function in IT2FIS

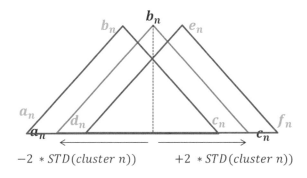

$$-2 * STD(cluster\ n)) \qquad +2 * STD(cluster\ n))$$

3.4.2.2 Gaussian Membership Functions

In Sect. 3.4.1.2 we explained the design of Gaussian Membership Functions for T1FIS with only two-points for that. Gaussian Membership Functions for IT2FIS is designing with three points, which are: σ, c_1, and c_2.

It is important to mention that the Gaussian Membership Functions for IT2FIS in this book, they are Gaussian Membership Functions with uncertainty on the mean, and that is to say, the error parameter is added in the mean of the function. In Table 3.9 the calculus for each point is shown.

Table 3.9 Calculation of the points in Gaussian Membership Functions for IT2FIS

Membership function	Formula
$Membership Function_1$	$\sigma_1 = STD(cluster\ 1)$
	$c_{a1} = centroide(cluster\ 1) - error$
	$c_{b1} = centroide(cluster\ 1) - error$
$Membership Function_2$	$\sigma_2 = STD(cluster\ 2)$
	$c_{a2} = centroide(cluster\ 2) - error$
	$c_{b2} = centroide(cluster\ 2) - error$
...	$\sigma_n = STD(cluster\ n)$
$Membership Function_n$	$c_{an} = centroide(cluster\ n) - error$
	$c_{bn} = centroide(cluster\ n) - error$

Fig. 3.27 Graphic representation of Gaussian Membership Function in IT2FIS

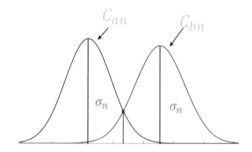

In Fig. 3.27, we present a general graphic representation of a Gaussian Membership function for IT2FIS. C_{an} and C_{bn} represents means with uncertainty and σ_n represent the standard deviation.

3.4.2.3 Trapezoidal Membership Functions

The calculus for each point in a Trapezoidal Membership function for IT2FIS is shown in Table 3.10.

Figure 3.28 shows a graphic representation of a Trapezoidal Membership function for IT2FIS.

As we can see, in this Membership function we need eight points: a, b, c, d, e, f, g and h to designing a Trapezoidal Membership function for IT2FIS.

Table 3.10 Calculus of the points in Trapezoidal Membership Functions for IT2FIS

Membership function	Formula
	$a_1 = min(cluster1) - error$
	$b_1 = Cuartil\,1(cluster\,1) - error$
	$c_1 = Cuartil\,3(cluster\,1) - error$
$MembershipFunction_1$	$d_1 = max(cluster1) - error$
	$e_1 = min(cluster1) + error$
	$f_1 = Cuartil\,1(cluster\,1) + error$
	$g_1 = Cuartil\,3(cluster\,1) + error$
	$h_1 = max(cluster\,1) + error$
	$a_2 = min(cluster\,2) - error$
	$b_2 = Cuartil\,1(cluster\,2) - error$
	$c_2 = Cuartil\,3(cluster\,2) - error$
$MembershipFunction_2$	$d_2 = max(cluster\,2) - error$
	$e_2 = min(cluster\,2) + error$
	$f_2 = Cuartil\,1(cluster\,2) + error$
	$g_2 = Cuartil\,3(cluster\,2) + error$
	$h_2 = max(cluster\,2) + error$
…	$a_n = min(cluster\,n) - error$
	$b_n = Cuartil\,1(cluster\,n) - error$
	$c_n = Cuartil\,3(cluster\,n) - error$
$MembershipFunction_n$	$d_n = max(cluster\,n) - error$
	$e_n = min(cluster\,n) + error$
	$f_n = Cuartil\,1(cluster\,n) + error$
	$g_n = Cuartil\,3(cluster\,n) + error$
	$h_n = max(cluster\,n) + error$

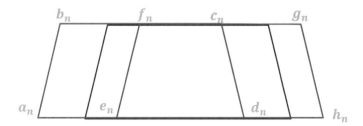

Fig. 3.28 Design of a Trapezoidal Membership function in IT2FIS

References

1. Barraza, J., Rodríguez, L., Castillo, O., Melin, P., Valdez, F.: A new hybridization approach between the fireworks algorithm and Grey Wolf optimizer algorithm. J. Optim. Res. Article **2018** (18 pages). Article ID 6495362 (2018)
2. Castro, J.R., Castillo, O., Melin, P., Rodriguez, A.: A Hybrid Learning Algorithm for Interval Type-2 Fuzzy Neural Networks: The Case of Time Series Prediction, vol. 15a, pp. 363–386. Springer, Berlin Heidelberg (2008)
3. Hagras, H.: Comments on dynamical optimal training for interval type-2 fuzzy neural network (T2FNN). IEEE Trans. Syst. Man Cybern. Part B **36**(5), 1206–1209 (2006)
4. Jang, J.S.R., Sun, C.T., Mizutani, E.: Neuro-Fuzzy and Soft Computing. Prentice-Hall, New York (1997)
5. Melin, P., Soto, J., Castillo, O., Soria, J.: A new approach for time series prediction using ensembles of ANFIS models. Experts Syst. Appl. Elsevier **39**(3), 3494–3506 (2012)
6. Sharkey, A.: Combining Artificial Neural Nets: Ensemble and Modular Multi-Net Systems. Springer, London (1999)
7. Soto, J., Melin, P., Castillo, O.: Time series prediction using ensembles of ANFIS models with genetic optimization of interval type-2 and type-1 fuzzy integrators. Int. J. Hybrid Intell. Syst. **11**(3), 211–226 (2014)
8. Wei, L.Y., Cheng, C.H.: A hybrid recurrent neural networks model based on synthesis features to forecast the Taiwan Stock market. Int. J. Innov. Comput. Inf. Control **8**(8), 5559–5571 (2012)
9. Xue, J., Xu, Z., Watada, J.: Building an integrated hybrid model for short-term and mid-term load forecasting with genetic optimization. Int. J. Innov. Comput. Inf. Control **8**(10), 7381–7391 (2012)
10. Daugman, J.: Statistical richness of visual phase information: update on recognizing persons by iris patterns. Int. J. Comput. Vision **45**(1), 25–38 (2001)
11. Khaw, P.: Iris recognition technology for improved authentication. In: Sala de Lectura de Seguridad de la Información, pp. 1–17. SANS Institute (2002)
12. Barraza, J., Melin, P., Valdez, F., González, C.I., Castillo, O.: Iterative Fireworks Algorithm with Fuzzy Coefficients, pp. 1–6. FUZZ-IEEE 2017, Naples, Italy
13. Barraza, J., Melin, P., Valdez, F.: Fuzzy FWA with Dynamic Adaptation of Parameters, pp. 4053–4060. IEEE CEC 2016, Vancouver, Canada
14. Barraza, J., Melin, P., Valdez, F., González, C.I.: Fireworks algorithm (FWA) with adaptation of parameters using fuzzy logic. In: Nature-Inspired Design of Hybrid Intelligent Systems, pp. 313–327 (2017)
15. Tan, Y., Zhu, Y.: Fireworks algorithm for optimization. In: Advances in Swarm Intelligence, vol. 6145, pp. 355–364. Springer, Berlin Heidelberg (2010)
16. Davis, L.: Handbook of Genetic Algorithms. Van Nostrand Reinhold (1991).
17. Deb, K.: A Population-Based Algorithm-Generator for Real-Parameter Optimization. Springer, Heidelberg (2005)
18. Telescaa, L., Bernardib, M., Rovellib, C.: Intra-cluster and inter-cluster time correlations in lightning sequences. Phys. A **356**, 655–661 (2005)
19. Escalante, H.J., Montes, M., Sucar, L.E.: Particle swarm model selection. J. Mach. Learn. Res. **10**, 405–440 (2009)
20. Eberhart, R., Shi, Y., Kennedy, J.: Swarm Intelligence. San Mateo, California, Morgan Kaufmann (2001).
21. Barraza, J., Melin, P., Valdez, F., González, C.I.: Fuzzy Fireworks Algorithm Based on a Sparks Dispersion Measure. Algorithms 10(3), 83 (2017)
22. Liu, J., Zheng, S., Tan, Y.: The improvement on controlling exploration and exploitation of firework algorithm. In: Advances in Swarm Intelligence, pp. 11–23. Springer (2013)

23. Črepinšek, M., Liu, S.H., Mernik, M.: Exploration and exploitation in evolutionary algorithms: a survey. ACM Comput. Surv. **45**(3), 35:32 (2013).
24. Barraza, J., Melin, P., Valdez, F., González, C.I.: Fireworks algorithm (FWA) with adaptation of parameters using interval type-2 fuzzy logic system. In: Intuitionistic and Type-2 Fuzzy Logic Enhancements in Neural and Optimization Algorithms, pp. 35–47 (2020)

Chapter 4
Simulation Studies of the Hybrid of Fuzzy Fireworks and Competitive Neural Networks

In this chapter we present results obtained throughout of this book. The results are partitioned by methods, that is to say, firstly, the results of the optimization methods are presented, and secondly, the results of the classification methods with fuzzy classifiers are shown.

4.1 Optimization Methods

This section presents the simulation and test results in the Optimization methods. We have developed three variants of the optimization methods, Fireworks Algorithm for Clustering (FWAC), Fuzzy Fireworks Algorithm for Clustering (FFWAC) and Interval Type 2 Fuzzy Fireworks Algorithm for Clustering (F2FWAC). The algorithm variants for clustering has two metrics for determining the maximum number of clusters (Square root of N and Sturges Law) and two metrics for evaluating the clusters quality (Intra-cluster and Inter-cluster) with minimum and maximum distance, thus, all possible combinations for obtaining results, in the following sections are presented.

4.1.1 Fireworks Algorithm for Clustering (FWAC)

Fireworks Algorithm for clustering is the first variation of the FWA adapted for optimizing the optimal number of clusters. The parameters for their execution are presented in Table 4.1.

4.1.1.1 Iris Dataset

Table 4.2 shows the results of FWAC for Iris dataset using Intra-cluster, and, Table 4.3 shows the results using Inter-cluster as metric.

© The Author(s), under exclusive license to Springer Nature Switzerland AG 2023
F. Valdez et al., *Hybrid Competitive Learning Method Using the Fireworks Algorithm and Artificial Neural Networks*, SpringerBriefs in Computational Intelligence, https://doi.org/10.1007/978-3-031-47712-6_4

Table 4.1 Parameters of FWAC

Parameter	Value
Fireworks (X_i)	5
Amplitude coefficient ($\widehat{A}$)	40
Spark coefficient (m)	50
Dimensions (d)	Features data
Function evaluations	15,000

The best results obtained were using Sturges Law with minimum distance for Intra-cluster and, with maximum distance for Inter-cluster. Such results were marked with bold on the mean and standard deviation (STD).

4.1.1.2   Wine Dataset

Tables 4.4 and 4.5 show the results of FWAC for Iris dataset using Intra-cluster and Inter-cluster, respectively.

For wine dataset, the best results were using Inter-cluster as cluster validation, and Sturges Law with maximum distance; results were marked in bold.

4.1.1.3   Breast Cancer Wisconsin Diagnostic (WDBC)

Table 4.6 shows the performance of FWAC for WDBC dataset using Intra-cluster as metric cluster validation.

Table 4.7 shows the performance of FWAC for WDBC dataset using Inter-cluster as metric cluster validation. The best result is marked in bold, was using Inter-cluster cluster validation and Sturges Law with maximum distance.

4.1.2   Type 1 Fuzzy Logic Fireworks Algorithm for Clustering (FFWAC)

Type 1 Fireworks Algorithm for clustering is the second variation of the FWA adapted for optimizing the optimal number of clusters. The parameters for their execution are presented in Table 4.8.

The results obtained with FFWAC for Iris dataset using Intra-cluster and Inter-cluster are shown in Tables 4.9 and 4.10 with FFWAC-I, Tables 4.11 and 4.12 with FFWAC-II, and, Tables 4.13 and 4.14 (FFWAC-III), respectively.

Tables 4.15, 4.17 and 4.19 show the results using Intra-cluster; Tables 4.16, 4.18 and 4.20 using Inter-cluster of the FFWAC-I, FFWAC-II and FFWAC-III, respectively.

Table 4.2 Summary of the CEC 15 learning-based benchmark suite

Distance	Distance minimum				Distance maximum			
Metric	$\sqrt{n}$		*Sturges*		$\sqrt{n}$		*Sturges*	
Fitness/number	Fit	K	Fit	K	Fit	K	Fit	K
1	0.572	4	0.372	2	4.811	8	5.398	6
2	0.660	7	0.575	5	4.787	3	4.790	5
3	0.612	7	0.562	3	4.967	6	4.975	8
4	0.614	7	0.383	2	4.756	8	4.903	2
5	0.625	2	0.509	4	5.043	9	4.200	4
6	0.530	4	0.539	2	4.721	9	4.500	5
7	0.583	7	0.521	5	4.539	6	4.955	7
8	0.505	6	0.604	6	4.472	4	4.825	2
9	0.488	8	0.543	2	4.805	5	4.844	7
10	0.621	7	0.421	3	5.529	10	4.755	7
11	0.486	3	0.490	3	4.651	6	4.663	5
12	0.501	2	0.432	2	4.834	11	4.771	7
13	0.521	4	0.437	5	4.836	10	4.764	4
14	0.530	7	0.495	2	5.461	7	4.835	7
15	0.652	5	0.567	2	5.408	3	5.406	6
16	0.480	6	0.461	3	4.689	5	4.757	6
17	0.611	6	0.654	6	4.723	8	4.834	2
18	0.675	7	0.617	6	4.687	6	4.878	4
19	0.497	8	0.543	6	4.659	12	4.710	5
20	0.513	2	0.507	3	5.007	6	4.777	6
21	0.465	6	0.480	2	4.829	7	4.875	7
22	0.587	9	0.506	2	4.993	11	4.821	7
23	0.524	3	0.456	3	4.728	11	4.765	3
24	0.551	7	0.515	5	4.602	4	4.815	6
25	0.406	2	0.431	2	4.821	6	5.234	2
26	0.516	3	0.527	7	4.462	9	4.510	4
27	0.604	6	0.545	3	4.828	7	4.623	8
28	0.512	5	0.546	4	4.097	5	5.068	3
29	0.440	3	0.483	2	4.826	4	5.186	2
30	0.160	2	0.579	3	4.810	9	4.552	3
31	0.535	4	0.515	2	5.745	2	4.776	7
Mean	5.13		**3.45**		7.00		5.06	
STD	2.13		**1.59**		2.66		1.97	

Table 4.3 Results of FWAC for Iris dataset using Inter-cluster

Distance	Minimum				Maximum			
Metric	$\sqrt{n}$		*Sturges*		$\sqrt{n}$		*Sturges*	
fitness/number	Fit	K	Fit	K	Fit	K	Fit	K
1	0.012	4	0.031	3	23.060	6	16.403	3
2	0.165	5	0.002	2	34.922	4	23.556	3
3	0.203	3	0.052	3	25.569	3	11.982	2
4	0.083	6	0.075	3	25.201	3	18.060	2
5	0.000	2	0.002	2	29.473	3	15.454	2
6	0.067	3	0.000	2	30.205	5	20.185	4
7	1.058	5	0.310	3	30.465	4	18.345	2
8	0.081	4	0.206	4	25.780	5	11.546	2
9	0.109	4	0.042	3	20.127	2	20.707	4
10	0.021	3	0.048	4	34.036	2	13.271	5
11	0.055	3	0.056	6	33.740	3	17.848	3
12	0.071	5	0.076	3	28.811	6	12.962	3
13	0.000	2	0.000	2	26.430	3	17.144	4
14	0.202	4	0.135	3	28.957	3	16.746	3
15	0.390	5	0.032	2	27.687	3	14.076	2
16	0.000	2	0.016	3	27.344	8	19.730	2
17	0.380	5	0.282	4	22.436	4	16.810	2
18	0.355	7	0.000	3	24.065	5	22.545	2
19	0.478	5	0.000	2	27.351	3	13.942	3
20	0.002	12	0.000	2	28.046	3	12.683	2
21	0.004	2	0.001	2	23.983	2	16.020	6
22	0.028	3	0.028	4	27.646	8	15.192	3
23	0.050	3	0.004	2	32.210	3	15.100	3
24	0.083	3	0.005	2	24.371	3	14.802	2
25	0.043	5	0.005	3	28.472	6	16.595	3
26	0.003	6	0.001	2	29.173	3	18.646	5
27	0.007	2	0.001	2	24.035	4	20.218	5
28	0.231	4	0.049	3	29.355	4	14.928	2
29	0.003	2	0.009	2	29.617	3	13.542	5
30	0.023	3	0.188	4	26.738	6	21.889	2
31	0.041	3	0.001	2	25.720	7	19.827	5
Mean	4.03		**2.81**		4.10		**3.10**	
STD	2.01		**0.95**		1.68		**1.22**	

Table 4.4 Results of FWAC for wine dataset using Intra-cluster

Distance	Minimum				Maximum			
Metric	$\sqrt{n}$		*Sturges*		$\sqrt{n}$		*Sturges*	
fitness/ number	Fit	K	Fit	K	Fit	K	Fit	K
1	13.813	10	20.749	3	203.627	10	322.15	3
2	17.632	12	20.976	7	165.552	9	521.05	5
3	14.712	12	16.680	2	442.732	8	538.67	6
4	18.961	10	17.783	8	343.646	12	254.93	5
5	15.915	11	23.096	2	339.952	8	300.83	7
6	18.876	6	17.556	8	271.779	7	288.04	7
7	14.591	13	16.647	5	508.375	6	344.53	5
8	19.019	4	12.800	2	297.436	12	336.27	4
9	17.743	5	19.357	3	396.649	10	473.50	5
10	15.697	12	19.865	8	245.152	11	447.79	2
11	18.518	9	17.612	8	195.036	10	369.40	6
12	15.901	10	20.435	6	283.212	11	427.15	5
13	19.848	10	18.233	7	302.181	10	556.49	2
14	16.118	9	18.096	8	422.345	6	282.52	8
15	12.751	6	20.673	4	342.172	4	315.95	6
16	15.509	3	18.464	4	420.096	8	261.97	7
17	15.278	12	16.420	7	307.424	11	366.46	3
18	15.407	9	18.834	5	769.650	13	323.08	3
19	14.914	9	19.789	6	305.126	7	560.37	5
20	20.631	8	15.920	7	276.762	6	508.86	6
21	21.669	4	16.402	5	233.420	7	551.42	6
22	19.383	3	18.249	4	301.662	9	320.51	5
23	16.820	11	17.637	7	353.287	11	534.76	7
24	16.432	8	20.302	7	317.630	12	221.00	5
25	19.332	11	19.337	6	196.663	12	337.38	5
26	15.311	13	18.553	8	645.555	8	418.62	7
27	13.732	10	15.203	8	488.746	3	437.71	8
28	13.676	4	20.042	3	301.476	4	400.80	5
29	16.826	13	17.847	6	293.608	7	514.78	4
30	18.802	7	21.054	8	220.115	5	341.80	3
31	15.936	13	15.783	2	453.236	13	505.06	4
Mean	8.94		5.61		8.71		5.13	
STD	3.23		2.16		2.80		1.63	

Table 4.5 Results of FWAC for wine dataset using Inter-cluster

Distance	Minimum				Maximum			
Metric	$\sqrt{n}$		*Sturges*		$\sqrt{n}$		*Sturges*	
fitness/ number	Fit	K	Fit	K	Fit	K	Fit	K
1	21.019	7	8.677	2	3115.95	7	2078.58	8
2	2.720	2	1.676	2	4012.84	7	2459.49	5
3	9.024	4	6.042	4	3506.15	9	2278.81	4
4	4.800	5	0.583	2	2821.64	3	2133.63	3
5	7.974	3	4.026	2	3709.69	7	1801.12	3
6	5.669	2	14.446	3	3307.20	8	2079.16	3
7	24.217	6	34.895	8	2943.77	3	2214.34	4
8	1.115	3	2.082	2	3256.13	4	2612.66	2
9	23.672	3	17.892	5	2005.16	7	2530.85	6
10	0.960	2	8.387	3	3199.35	3	2219.86	2
11	17.347	3	3.338	2	2852.29	7	2130.86	6
12	10.573	4	0.746	2	3857.55	3	2200.11	3
13	13.804	5	5.701	3	3369.68	9	2007.04	3
14	21.705	4	20.778	7	1770.97	2	2379.21	4
15	23.282	5	0.503	2	3592.01	2	2551.73	3
16	12.289	9	21.882	5	2940.58	2	1903.03	2
17	0.602	2	3.360	4	3502.91	4	2542.92	4
18	1.478	2	0.728	2	3568.17	2	2174.09	4
19	1.245	2	0.842	4	3261.27	4	1998.95	3
20	12.641	8	0.511	3	3509.05	3	1968.61	3
21	25.596	4	15.612	3	2367.60	2	2007.20	5
22	3.823	2	18.957	3	3790.03	3	921.31	2
23	3.733	2	26.617	7	3683.28	5	2200.39	3
24	43.213	7	3.426	3	3526.61	5	1798.94	3
25	42.943	5	0.148	2	3449.51	2	1720.14	4
26	9.620	3	0.757	2	3173.76	4	2308.67	3
27	1.716	2	20.726	6	3335.13	3	1668.04	3
28	21.547	3	9.711	3	2784.56	4	1494.51	2
29	29.499	7	5.102	4	3653.68	3	2096.01	3
30	61.801	6	0.922	3	3519.81	5	1833.49	2
31	8.915	3	2.205	2	4199.65	3	2129.66	3
Mean	4.03		**3.39**		4.35		**3.48**	
STD	2.03		**1.69**		2.18		**1.36**	

Table 4.6 Results of FWAC for WDBC dataset using Intra-cluster

Distance	Minimum				Maximum			
Metric	$\sqrt{n}$		*Sturges*		$\sqrt{n}$		*Sturges*	
Fitness/ number	Fit	K	Fit	K	Fit	K	Fit	K
1	138.941	19	32.851	3	2439.49	2	1662.516	6
2	21.484	6	178.964	8	1821.03	22	1992.226	6
3	71.056	3	44.150	9	1391.14	13	2106.326	9
4	158.040	14	175.933	6	1622.74	17	1372.916	9
5	50.680	2	26.509	6	2313.85	2	2272.887	10
6	27.355	3	46.164	2	2053.42	22	2060.830	8
7	41.119	5	23.422	9	1097.58	20	1379.768	9
8	30.572	2	36.189	4	2601.10	12	1736.096	5
9	91.731	2	34.933	4	1676.17	8	1493.031	9
10	23.890	12	34.589	4	1201.14	20	1937.872	9
11	42.569	8	172.207	9	1269.59	12	1272.411	7
12	60.636	2	41.581	8	2137.23	24	2174.077	6
13	168.668	9	166.185	8	1956.71	23	1203.133	5
14	26.649	19	186.764	9	1041.77	12	3109.389	9
15	160.736	20	136.482	9	2471.43	13	1731.238	4
16	116.293	15	165.516	7	1212.45	17	1952.226	4
17	128.312	20	35.309	3	1522.30	18	2935.999	2
18	33.937	18	42.921	7	2011.06	19	3049.913	3
19	26.517	16	39.358	5	1678.36	8	2228.745	7
20	28.534	2	28.205	6	1595.05	12	1590.556	8
21	148.730	15	37.837	6	1184.26	13	1343.993	7
22	143.225	18	172.956	9	1238.44	15	2183.738	3
23	28.435	3	37.170	3	1195.82	20	1651.983	10
24	146.594	19	21.903	6	1769.40	11	3609.113	5
25	26.680	13	26.542	3	3024.86	2	1592.371	4
26	20.128	5	28.239	4	1209.43	12	2010.475	8
27	152.869	12	135.973	9	1611.25	15	1649.205	2
28	158.233	20	35.859	8	1221.23	16	2121.278	8
29	24.542	5	32.788	2	3124.71	18	1312.652	9
30	138.364	19	202.858	6	2118.73	21	1639.809	10
31	148.922	18	29.603	9	1151.74	24	1223.436	6
Mean	11.10		6.16		14.94		6.68	
STD	6.98		2.41		6.21		2.45	

Table 4.7 Results of FWAC for WDBC dataset using Inter-cluster

Distance	Minimum				Maximum			
Metric	$\sqrt{n}$		*Sturges*		$\sqrt{n}$		*Sturges*	
Fitness/ number	Fit	K	Fit	K	Fit	K	Fit	K
1	27.814	8	0.9716	2	27,495.32	6	12,052.13	5
2	640.142	8	3.0920	2	29,935.93	17	11,427.10	3
3	54.317	7	1.6257	2	29,008.89	6	10,350.20	2
4	150.826	20	6.6486	2	31,416.82	15	9420.97	4
5	191.993	4	33.5232	3	26,760.55	3	11,256.34	8
6	340.231	7	24.3943	9	28,394.47	11	7498.70	5
7	0.152	2	188.3866	3	28,418.68	12	10,464.32	3
8	91.968	4	25.6134	3	22,652.57	11	11,989.93	3
9	845.529	22	19.7025	3	28,330.59	3	10,879.28	3
10	84.978	3	64.9797	3	26,955.89	20	13,268.61	3
11	165.744	3	234.1786	6	23,487.71	2	11,543.95	2
12	0.577	2	29.2372	2	26,438.98	19	12,569.04	5
13	544.309	14	32.9347	4	28,947.25	15	11,798.71	3
14	11.977	3	215.0969	4	31,470.65	10	11,710.65	6
15	134.237	5	69.3285	3	27,324.51	7	8831.42	5
16	3.090	2	45.4879	3	25,427.87	11	14,085.28	5
17	46.881	3	97.8342	7	31,781.05	8	10,460.92	3
18	187.804	5	12.5064	7	28,612.35	13	12,880.41	3
19	261.203	7	112.6738	5	25,226.51	6	12,248.58	6
20	15.522	3	7.4007	9	19,133.16	17	10,899.73	7
21	66.162	4	285.5903	8	24,062.08	4	11,709.93	3
22	13.001	5	173.1757	9	17,451.82	2	12,963.23	2
23	110.898	5	190.4979	4	31,327.74	13	10,517.35	3
24	72.094	3	63.0232	3	28,224.43	6	10,297.62	2
25	1904.225	20	23.4938	2	26,203.69	3	14,493.02	3
26	51.254	4	9.3412	2	18,956.74	5	13,131.28	3
27	3.294	2	93.6991	5	29,075.07	18	11,256.29	6
28	521.624	12	166.3302	9	22,366.56	2	10,430.67	5
29	374.057	8	2.4696	2	28,539.86	9	13,589.44	2
30	248.425	7	152.0142	4	24,706.26	18	12,740.90	3
31	351.429	8	82.7344	8	23,897.07	6	13,659.85	4
Mean	6.77		4.45		9.61		**3.87**	
STD	5.44		2.51		5.69		**1.59**	

Table 4.8 Parameters of FFWAC

Parameter	Value
Fireworks (X_i)	5
Amplitude coefficient ($\widehat{A}$)	[40, 2]
Spark coefficient (m)	50
Dimensions (d)	Features data
Function evaluations	15,000
Membership functions (Type 1 fuzzy logic)	• Triangular • Gaussian • Trapezoidal

For WDBC dataset, Tables 4.21, 4.23 and 4.25 show the result using Intra-cluster of FFWAC-I, FFWAC-II and FFWAC-III, respectively; and, Tables 4.22, 4.24 and 4.26 show the results using Inter-cluster of FFWAC-I, FFWAC-II and FFWAC-III, respectively.

4.1.2.1 Iris Dataset

For Iris dataset, the best results obtained by the FFWAC-I was using maximum distance in Sturges Law with Intra and Inter-cluster.

The best results obtained of the FFWAC-II were using maximum distance in Sturges Law with Intra and Inter-cluster for the Iris dataset.

As the same of FFWAC-I and FFWAC-II, the best results of FFWAC-III were using maximum distance in Sturges Law with Intra and Inter-cluster for the Iris dataset.

4.1.2.2 Wine Dataset

The best results of FFWAC-I for the wine dataset was using the minimum and maximum distance with the Sturges Law in Inter-cluster.

The best performance of FFWAC-II for wine dataset is using Inter-cluster as cluster validation.

The best results of the FFWAC-III obtained for wine dataset is using the Inter-cluster as cluster validation.

4.1.2.3 Breast Cancer Wisconsin Diagnostic (WDBC)

The results of the FFWAC-I for WDBC dataset was not satisfactory.

The best result obtained of the FFWAC-II performance for WDBC dataset is using Inter-cluster with maximum distance in Sturges Law.

Table 4.9 Results of FFWAC-I for Iris dataset using Intra-cluster

Distance	Minimum				Maximum			
Metric	$\sqrt{n}$		$Sturges$		$\sqrt{n}$		$Sturges$	
Fitness/number	Fit	K	Fit	K	Fit	K	Fit	K
1	0.559	3	0.606	4	19.809	2	21.007	3
2	0.507	4	0.600	5	31.131	11	16.49	4
3	0.613	8	0.403	2	22.055	3	18.298	4
4	0.610	6	0.559	4	32.066	3	18.497	2
5	0.406	3	0.520	6	26.789	9	16.159	7
6	0.522	11	0.521	6	29.164	8	22.013	2
7	0.420	4	0.545	6	30.421	9	17.352	5
8	0.534	6	0.535	3	24.882	4	19.423	3
9	0.566	10	0.488	2	23.78	6	19.188	3
10	0.535	3	0.508	6	15.172	2	19.241	5
11	0.539	3	0.422	4	29.354	3	11.15	3
12	0.488	3	0.477	6	33.218	2	10.393	2
13	0.611	10	0.428	2	24.666	4	19.813	5
14	0.609	3	0.516	3	26.193	2	17.01	5
15	0.541	10	0.589	6	27.784	3	20.165	3
16	0.514	5	0.513	4	23.507	2	21.942	3
17	0.348	2	0.499	7	27.733	3	13.451	3
18	0.573	10	0.595	6	19.261	3	17.1	5
19	0.524	3	0.520	2	19.088	5	13.444	2
20	0.512	10	0.407	2	20.895	3	6.534	2
21	0.501	12	0.465	2	28.63	12	19.304	2
22	0.504	9	0.397	3	26.505	2	17.523	2
23	0.503	3	0.485	3	26.169	2	17.845	2
24	0.468	6	0.361	2	31.405	5	22.345	3
25	0.589	11	0.486	2	17.772	4	20.446	3
26	0.409	2	0.554	4	22.872	9	18.166	2
27	0.543	4	0.570	5	24.752	4	18.815	2
28	0.574	9	0.559	4	22.31	3	17.925	3
29	0.582	10	0.558	6	21.682	4	17.838	3
30	0.542	12	0.478	2	25.896	5	13.118	2
31	0.518	4	0.366	3	24.716	3	17.962	3
Mean	6.42		3.94		4.52		**3.16**	
STD	3.44		1.67		2.83		**1.27**	

Table 4.10 Results of FFWAC-I for Iris dataset using Inter-cluster

Distance	Minimum				Maximum			
Metric	$\sqrt{n}$		$Sturges$		$\sqrt{n}$		$Sturges$	
Fitness/number	Fit	K	Fit	K	Fit	K	Fit	K
1	0.025	9	0.096	3	20.13	3	17.641	3
2	0.073	3	0.001	2	22.94	3	16.606	4
3	0.005	2	0.006	2	23.64	7	17.113	3
4	0.114	3	0.049	3	25.93	2	14.187	4
5	0.076	4	0.000	8	26.31	3	12.812	4
6	0.308	10	0.002	2	28.87	6	16.203	3
7	0.409	5	0.004	3	22.84	3	21.038	2
8	0.267	8	0.020	3	18.55	3	13.863	4
9	0.021	4	0.015	2	31.5	4	21.679	3
10	0.335	4	0.285	6	23.88	6	20.593	8
11	0.263	4	0.045	2	29.97	12	15.377	2
12	0.056	3	0.385	7	25.7	10	11.994	2
13	0.060	5	0.005	2	23.73	3	17.659	3
14	0.008	2	0.014	8	27.14	3	17.028	5
15	0.200	3	0.170	3	28.82	4	21.077	4
16	0.002	2	0.295	6	17.43	3	17.233	2
17	0.000	2	0.153	5	26.2	4	19.031	5
18	0.050	3	0.037	3	21.96	3	7.458	3
19	0.107	3	0.009	3	26.57	3	15.024	3
20	0.031	3	0.118	3	28.29	2	19.543	3
21	0.003	3	0.165	3	22.15	3	20.396	2
22	0.200	6	0.028	2	24.86	4	19.457	2
23	0.212	10	0.113	3	26.86	3	9.292	2
24	0.018	2	0.061	4	23.04	5	17.106	3
25	0.470	5	0.368	4	18.32	4	17.297	2
26	0.217	3	0.048	4	29.59	3	16.484	4
27	0.128	4	0.277	4	29.41	8	22.298	2
28	0.000	2	0.043	3	21.48	2	15.898	2
29	0.632	12	0.016	3	24.47	4	16.478	5
30	0.978	9	0.025	6	24.65	6	14.860	4
31	0.062	3	0.163	6	29.38	4	18.758	5
Mean	4.55		3.80		4.29		**3.32**	
STD	2.79		1.79		2.31		**1.35**	

Table 4.11 Results of FFWAC-II for Iris dataset using Intra-cluster

Distance	Minimum				Maximum			
Metric	$\sqrt{n}$		*Sturges*		$\sqrt{n}$		*Sturges*	
Fitness/number	Fit	K	Fit	K	Fit	K	Fit	K
1	0.556	7	0.581	2	25.608	6	10.751	3
2	0.591	11	0.504	5	31.37	9	18.557	3
3	0.558	9	0.577	8	26.25	3	13.376	3
4	0.562	4	0.436	6	17.389	3	12.986	2
5	0.513	4	0.493	7	18.827	2	16.333	3
6	0.572	10	0.572	5	30.079	2	16.429	2
7	0.486	8	0.473	6	22.959	4	19.14	4
8	0.508	4	0.577	3	29.257	8	17.69	2
9	0.578	8	0.533	3	26.428	6	19.455	4
10	0.522	2	0.555	6	25.253	2	18.167	4
11	0.416	6	0.527	4	19.03	3	22.729	3
12	0.438	6	0.565	8	33.198	2	17.446	3
13	0.529	5	0.469	5	27.657	3	18.944	3
14	0.550	8	0.570	7	27.988	4	15.753	2
15	0.546	8	0.494	3	26.588	3	18.109	3
16	0.455	2	0.605	7	26.094	3	20.201	4
17	0.499	11	0.574	8	23.296	3	15.801	4
18	0.554	5	0.602	3	18.966	2	11.763	2
19	0.592	9	0.589	5	25.79	2	19.77	4
20	0.610	8	0.544	7	24.195	7	13.7	4
21	0.548	6	0.484	3	24.457	2	15.174	5
22	0.530	11	0.560	6	19.695	4	13.094	3
23	0.501	8	0.550	6	32.931	2	18.616	6
24	0.450	3	0.513	4	17.582	3	18.821	3
25	0.612	5	0.596	2	22.197	2	16.428	2
26	0.485	7	0.393	3	25.109	4	16.417	6
27	0.552	11	0.505	7	22.177	5	15.28	4
28	0.597	12	0.553	3	27.31	6	19.522	2
29	0.544	9	0.439	3	19.897	5	16.459	2
30	0.619	8	0.498	2	22.559	3	21.71	3
31	0.562	9	0.532	6	18.927	3	18.803	2
Mean	7.23		4.94		3.74		**3.23**	
STD	2.75		1.95		1.9		**1.12**	

Table 4.12 Results of FFWAC-II for Iris dataset using Inter-cluster

Distance	Minimum				Maximum			
Metric	$\sqrt{n}$		Sturges		$\sqrt{n}$		Sturges	
Fitness/number	Fit	K	Fit	K	Fit	K	Fit	K
1	0.320	7	0.11	4	27.56	5	22.02	5
2	0.000	2	0.44	7	18.85	3	13.91	2
3	0.150	4	0.00	2	29.78	6	15.60	2
4	0.100	4	0.00	2	32.31	5	20.05	4
5	1.130	10	0.13	3	24.57	6	20.81	3
6	0.170	5	0.00	2	27.21	3	20.49	3
7	1.210	9	0.14	7	26.83	2	14.93	2
8	0.120	9	0.02	2	27.84	2	15.40	4
9	0.120	7	0.00	2	24.39	3	20.96	3
10	0.000	2	0.03	3	25.72	2	8.67	2
11	0.070	3	0.01	2	32.46	2	14.96	3
12	0.000	2	0.01	2	22.51	2	16.13	3
13	0.120	5	0.59	7	27.87	3	15.15	3
14	0.050	3	0.39	5	19.57	3	17.31	2
15	0.580	7	0.06	3	31.14	3	20.37	2
16	0.330	7	0.01	3	22.03	2	15.28	3
17	0.500	11	0.10	3	19.03	2	12.64	3
18	0.230	4	0.06	3	25.02	2	20.39	3
19	0.070	3	0.00	2	30.54	4	13.57	3
20	0.600	6	0.10	4	29.87	3	16.03	2
21	0.540	5	0.02	3	26.70	8	21.01	3
22	0.010	2	0.00	2	21.59	4	18.40	3
23	0.210	3	0.00	2	25.53	6	17.79	3
24	0.200	4	0.05	6	24.43	5	11.64	3
25	0.070	3	0.01	4	26.84	2	13.89	2
26	0.050	3	0.00	2	27.49	3	19.97	3
27	0.000	2	0.02	3	20.46	6	16.36	3
28	0.190	3	0.00	2	24.95	2	13.78	4
29	0.010	2	0.00	3	26.14	7	17.87	3
30	0.130	8	0.04	3	15.95	2	18.27	4
31	0.000	2	0.02	5	27.28	3	13.74	3
Mean	4.74		**3.32**		3.58		**2.94**	
STD	2.66		**1.60**		1.75		**0.73**	

Table 4.13 Results of FFWAC-III for Iris dataset using Intra-cluster

Distance	Minimum				Maximum			
Metric	$\sqrt{n}$		*Sturges*		$\sqrt{n}$		*Sturges*	
Fitness/number	Fit	K	Fit	K	Fit	K	Fit	K
1	0.565	2	0.485	6	19.81	2	21.01	3
2	0.514	10	0.533	3	31.13	11	16.49	4
3	0.574	9	0.513	7	22.06	3	18.30	4
4	0.445	3	0.453	4	32.07	3	18.50	2
5	0.563	11	0.137	2	26.79	9	16.16	7
6	0.477	5	0.593	6	29.16	8	22.01	2
7	0.570	11	0.555	8	30.42	9	17.35	5
8	0.496	7	0.498	4	24.88	4	19.42	3
9	0.544	9	0.570	4	23.78	6	19.19	3
10	0.536	7	0.496	3	15.17	2	19.24	5
11	0.592	3	0.590	2	29.35	3	11.15	3
12	0.537	6	0.333	2	33.22	2	10.39	2
13	0.532	10	0.553	3	24.67	4	19.81	5
14	0.572	9	0.528	6	26.19	2	17.01	5
15	0.484	12	0.534	7	27.78	3	20.17	3
16	0.614	11	0.459	5	23.51	2	21.94	3
17	0.460	6	0.177	3	27.73	3	13.45	3
18	0.557	10	0.391	6	19.26	3	17.10	5
19	0.528	4	0.535	3	19.09	5	13.44	2
20	0.523	5	0.555	5	20.90	3	6.53	2
21	0.585	6	0.505	4	28.63	12	19.30	2
22	0.506	3	0.509	3	26.51	2	17.52	2
23	0.578	8	0.581	5	26.17	2	17.85	2
24	0.512	5	0.471	3	31.41	5	22.35	3
25	0.565	5	0.540	7	17.77	4	20.45	3
26	0.614	6	0.561	6	22.87	9	18.17	2
27	0.404	2	0.601	2	24.75	4	18.82	2
28	0.529	8	0.505	3	22.31	3	17.93	3
29	0.612	11	0.299	3	21.68	4	17.84	3
30	0.597	5	0.529	4	25.90	5	13.12	2
31	0.517	10	0.431	5	24.72	3	17.96	3
Mean	7.06		4.32		4.52		**3.16**	
STD	3.00		1.72		2.83		**1.27**	

Table 4.14 Results of FFWAC-III for Iris dataset using Inter-cluster

Distance	Minimum				Maximum			
Metric	$\sqrt{n}$		*Sturges*		$\sqrt{n}$		*Sturges*	
Fitness/number	Fit	K	Fit	K	Fit	K	Fit	K
1	0.380	5	0.001	2	26.88	3	11.32	4
2	0.150	8	0.024	3	27.37	8	19.42	3
3	0.130	4	0.059	3	23.05	3	13.43	3
4	0.090	3	0.061	3	21.00	3	17.60	4
5	0.820	5	0.192	4	23.86	2	15.99	3
6	0.860	10	0.001	2	25.97	3	12.86	4
7	0.070	3	0.112	3	22.90	4	16.47	4
8	0.010	7	0.001	2	30.60	2	16.47	2
9	0.100	3	0.033	3	20.48	4	12.84	3
10	0.060	3	0.001	2	19.68	6	13.92	2
11	0.080	3	0.000	2	15.10	4	15.94	2
12	0.010	2	0.204	4	23.15	4	15.87	2
13	0.280	11	0.019	2	23.01	4	15.37	7
14	0.000	2	0.007	2	30.16	7	15.43	2
15	0.350	5	0.000	2	19.49	3	22.67	3
16	0.130	3	0.220	4	25.78	3	22.48	2
17	0.230	4	0.005	2	28.46	4	9.59	2
18	0.510	4	0.186	4	27.26	9	17.20	3
19	0.000	2	0.001	2	25.77	4	20.03	2
20	0.000	2	0.157	4	28.49	3	14.55	2
21	0.040	6	0.026	3	21.96	7	17.99	2
22	0.000	2	0.002	2	33.05	3	16.16	6
23	0.000	2	0.117	3	29.79	4	18.69	4
24	1.490	6	0.001	2	19.68	2	17.04	3
25	0.280	4	0.016	2	26.73	4	15.77	2
26	0.000	2	0.054	5	18.40	4	17.97	3
27	0.000	2	0.035	4	24.17	2	15.02	4
28	0.080	3	0.006	2	27.28	3	16.98	3
29	0.150	4	0.028	3	20.72	3	21.93	2
30	0.000	2	0.135	4	20.37	3	20.17	2
31	0.000	2	0.280	3	26.37	4	14.64	5
Mean	4		**2.84**		3.94		**3.06**	
STD	2.37		**0.90**		1.73		**1.26**	

Table 4.15 Results of FFWAC-I for wine dataset using Intra-cluster

Distance	Minimum				Maximum			
Metric	$\sqrt{n}$		*Sturges*		$\sqrt{n}$		*Sturges*	
Fitness/ number	Fit	K	Fit	K	Fit	K	Fit	K
1	13.964	3	16.347	3	3055.755	3	1917.26	6
2	19.145	3	18.815	6	2375.45	8	1620.32	5
3	12.316	3	15.526	3	3284.617	2	1913.82	7
4	19.056	9	19.767	3	2559.96	3	2504.59	4
5	16.706	11	19.606	7	3631.096	2	2174.11	5
6	15.972	10	15.322	5	3500.235	4	2052.91	3
7	18.254	5	18.652	8	2502.417	3	2215.93	3
8	15.521	9	18.118	7	2783.196	3	2250.81	8
9	13.216	5	13.241	2	3612.24	10	1892.66	3
10	14.234	10	16.514	6	3309.389	5	1964.61	6
11	16.204	8	16.819	8	3464.354	10	1925.1	3
12	19.291	8	21.181	5	3540.472	10	2002.4	5
13	16.660	9	12.367	4	3855.863	8	2052.59	5
14	16.257	13	13.843	7	2904.026	2	2099.07	4
15	16.760	12	19.824	5	3516.231	2	1852.67	3
16	15.520	13	17.749	8	3330.042	3	1789.47	2
17	17.972	11	20.226	7	3394.995	4	2264.95	4
18	15.862	9	17.186	8	2821.334	9	2051.65	2
19	15.341	11	17.774	7	2539.428	7	1829.79	2
20	19.814	9	18.608	8	3501.854	5	2054.9	6
21	15.985	11	14.467	3	2905.797	6	2535.29	4
22	17.466	7	19.083	6	3550.155	13	1648.52	4
23	20.134	12	22.555	4	3841.227	2	1816.7	3
24	15.947	11	23.257	4	3689.919	2	1796.2	2
25	20.028	3	17.694	7	2792.853	2	2268.36	4
26	12.975	5	15.872	7	3049.395	10	2038.34	4
27	15.469	10	22.164	2	3510.771	3	2054.09	4
28	15.928	4	12.978	3	2860.993	3	2179.78	2
29	16.128	11	16.748	8	3060.318	8	1627.96	5
30	16.032	11	20.239	4	2293.859	3	2476.7	2
31	18.741	10	15.661	5	3490.987	13	1460.72	3
Mean	8.58		5.48		5.42		**3.97**	
STD	3.16		2.00		3.5		**1.56**	

Table 4.16 Results of FFWAC-I for wine dataset using Inter-cluster

Distance	Minimum				Maximum			
Metric	$\sqrt{n}$		*Sturges*		$\sqrt{n}$		*Sturges*	
Fitness/number	Fit	K	Fit	K	Fit	K	Fit	K
1	13.87	3	1.13	3	2875.12	2	1986.64	3
2	6.37	3	0.01	2	2841.92	6	1602.72	3
3	14.00	5	14.92	7	3646.00	9	2100.11	6
4	11.49	8	19.43	4	3580.48	3	2342.40	4
5	18.13	6	5.83	3	3598.50	8	2511.35	7
6	14.57	3	0.01	2	3720.82	6	2192.29	2
7	62.52	11	26.87	3	3762.09	7	2010.39	4
8	2.81	2	4.84	2	2269.13	4	2052.74	5
9	12.60	5	1.35	2	3007.25	7	2290.77	2
10	3.89	4	37.34	4	2911.92	3	2433.41	6
11	42.55	9	6.45	8	3226.91	6	1780.55	6
12	85.87	11	0.07	2	3512.77	12	1847.62	2
13	22.47	10	0.41	2	1797.93	5	1589.64	3
14	27.70	8	1.61	2	2905.53	5	2221.54	3
15	2.01	2	30.04	4	3638.58	13	2008.43	2
16	32.09	3	3.27	3	2912.39	7	2076.39	2
17	16.73	3	1.33	4	3361.12	3	1437.23	4
18	39.71	6	17.15	4	2970.15	3	2050.64	3
19	0.05	2	13.51	7	2664.31	4	1832.82	3
20	41.18	4	4.65	4	3426.67	4	1952.90	2
21	3.12	3	0.46	2	3204.06	2	2455.80	2
22	84.96	10	1.15	2	2461.03	2	1910.03	2
23	6.63	3	0.57	2	3237.21	7	2473.62	3
24	3.47	3	1.98	2	2484.86	2	1590.13	2
25	8.58	7	29.14	3	3111.93	4	2270.51	7
26	15.40	10	11.47	4	2677.60	5	1800.58	3
27	71.36	6	2.04	2	3447.93	8	1151.73	3
28	10.96	4	2.77	7	3795.19	3	2388.66	3
29	13.93	3	5.58	4	2829.70	3	2334.54	4
30	18.73	4	13.24	7	3245.45	2	1397.58	4
31	41.37	4	12.91	3	2981.43	2	1381.58	2
Mean	5.32		**3.55**		5.06		**3.45**	
STD	2.90		**1.82**		2.87		**1.55**	

Table 4.17 Results of FFWAC-II for wine dataset using Intra-cluster

| Distance | Minimum | | | | Maximum | | | |
| Metric | $\sqrt{n}$ | | *Sturges* | | $\sqrt{n}$ | | *Sturges* | |
Fitness/number	Fit	K	Fit	K	Fit	K	Fit	K
1	19.783	4	16.519	8	3615.51	11	2154.052	7
2	12.864	3	20.079	6	2279.86	7	2247.041	7
3	16.177	12	20.884	2	3074.9	2	2006.91	2
4	13.479	13	21.007	8	3096.93	3	2484.954	5
5	17.108	3	13.490	7	3341.29	5	2070.595	2
6	16.551	12	19.830	3	3034.13	4	1907.402	4
7	18.570	8	13.895	5	3417.62	2	1907.233	3
8	16.189	12	14.439	6	3049.99	6	2311.832	6
9	17.214	10	18.922	6	2670.45	2	1931.181	6
10	13.877	13	19.782	6	3336.71	12	2346.768	6
11	14.538	13	20.398	7	2694.58	7	1989.55	2
12	17.313	7	19.241	7	3897.52	11	1616.894	7
13	13.138	11	19.668	4	2413.95	3	1985.909	3
14	15.588	12	17.025	7	3649.17	5	1838.26	5
15	16.782	11	16.800	4	2486.71	3	2105.235	3
16	18.528	10	20.225	7	3028.04	10	1785.49	4
17	17.775	9	18.330	4	2688.47	5	2583.775	3
18	16.071	12	15.902	3	2577.7	4	1931.534	2
19	20.017	11	13.887	7	2707.78	8	2012.451	3
20	19.830	11	16.442	8	2637.88	7	1874.606	5
21	17.151	11	18.547	3	2960.39	5	2114.851	3
22	15.089	13	18.026	7	3489.95	9	1750.354	3
23	19.302	7	15.368	2	3433.8	8	2242.449	2
24	18.103	10	17.601	5	3424.95	2	1829.844	2
25	15.513	9	16.938	6	3971.48	2	2288.226	3
26	18.615	10	17.252	7	3740.08	3	2261.689	2
27	20.992	4	22.568	3	2553.52	3	2131.281	3
28	15.823	11	16.743	8	3265.75	11	2322.629	4
29	15.724	9	22.649	7	3262.77	6	2150.369	2
30	18.075	7	19.752	6	3288.42	3	2172.983	8
31	19.536	7	14.883	3	2782.76	6	2289.337	5
Mean	9.52		5.55		5.65		3.94	
STD	2.99		1.91		3.1		1.82	

Table 4.18 Results of FFWAC-II for wine dataset using Inter-cluster

Distance	Minimum				Maximum			
Metric	$\sqrt{n}$		*Sturges*		$\sqrt{n}$		*Sturges*	
Fitness/number	Fit	K	Fit	K	Fit	K	Fit	K
1	29.860	3	1.84	2	3006.71	3	1493.33	2
2	13.350	4	2.26	2	3591.02	3	2303.09	5
3	58.900	6	0.68	3	3459.77	2	1716.86	5
4	18.310	8	0.08	2	2803.56	3	2370.47	4
5	56.180	4	11.48	3	3436.29	3	1903.09	4
6	16.210	6	29.34	7	2980.54	2	1858.48	5
7	2.100	2	31.78	4	3810.98	6	2164.53	3
8	8.730	2	0.51	2	3653.44	4	1812.46	7
9	75.600	3	0.88	2	3292.90	3	2324.51	4
10	3.630	3	9.83	3	2581.07	4	2155.87	6
11	33.690	6	27.06	5	3717.42	4	2184.76	2
12	0.530	2	1.43	2	3350.71	6	2513.87	4
13	0.510	2	1.46	2	2711.55	6	1700.95	3
14	1.730	3	44.11	6	2706.06	3	2408.84	3
15	32.570	4	46.29	6	3228.68	3	2344.19	2
16	2.280	2	0.39	2	2024.87	5	2173.71	4
17	3.590	6	6.93	3	2092.11	2	1723.55	2
18	23.670	6	9.35	3	3372.88	7	1992.01	5
19	23.830	3	4.06	5	3445.57	5	1870.85	4
20	2.300	2	0.41	2	3100.57	10	2591.49	2
21	23.400	11	51.64	4	3486.78	3	1797.57	5
22	15.350	10	0.47	2	3464.47	3	2148.25	2
23	18.580	4	26.65	3	3517.89	4	1812.27	3
24	5.440	4	2.51	2	2537.40	2	2277.40	5
25	4.890	4	0.75	2	3380.96	8	2403.81	3
26	3.920	3	0.31	2	3771.56	3	1982.78	2
27	34.040	9	11.48	4	3818.37	2	1480.65	3
28	0.000	2	0.13	2	3761.10	4	1992.06	2
29	6.910	8	9.44	7	2952.33	3	1736.77	3
30	28.750	3	0.07	5	3337.22	2	2134.62	2
31	6.790	5	54.96	7	3507.34	2	1384.10	2
Mean	4.52		**3.42**		**3.87**		**3.48**	
STD	2.51		**1.73**		**1.94**		**1.39**	

Table 4.19 Results of FFWAC-III for wine dataset using Intra-cluster

Distance	Minimum				Maximum			
Metric	$\sqrt{n}$		*Sturges*		$\sqrt{n}$		*Sturges*	
Fitness/ number	Fit	K	Fit	K	Fit	K	Fit	K
1	17.633	12	17.055	6	3439.66	9	2267.85	5
2	16.123	9	18.202	3	4000.42	2	2478.94	7
3	16.654	5	19.914	4	3099.42	2	1989.38	3
4	18.414	6	16.111	5	4128.33	7	2183.59	3
5	12.924	13	19.826	3	2142	2	2359.44	2
6	17.803	6	22.327	6	2613.07	2	2213.61	3
7	16.775	11	20.053	7	3926.39	4	2528.81	5
8	16.454	8	13.580	3	3982.17	2	1849.29	5
9	15.857	10	16.611	3	3113.58	12	2033.62	7
10	16.757	12	21.262	5	2924.51	8	2116.46	4
11	16.261	11	15.604	4	2927.97	7	2309.31	2
12	11.464	3	16.171	8	3961.75	2	1704.4	2
13	16.516	12	17.635	3	2780.62	3	1788.58	5
14	18.633	7	21.014	3	3925.56	2	2361.91	3
15	17.099	13	17.656	5	2848.6	8	2047.14	2
16	16.383	13	17.832	5	3919.11	11	1628.8	6
17	14.835	12	18.573	5	3666.71	5	1988.26	6
18	15.038	9	19.575	4	3236.99	3	1843.83	3
19	13.736	2	19.656	3	3246.78	7	1277.62	3
20	16.695	8	20.852	5	2721.11	2	2435.34	4
21	19.575	11	21.285	2	3473.32	10	2100.6	4
22	16.266	9	12.520	3	3294.69	2	2437.42	2
23	16.972	12	19.799	6	2448.79	9	1854.39	2
24	17.485	12	13.300	3	3656.52	5	2175.12	4
25	20.631	2	13.970	8	3016.29	2	1858.5	3
26	15.795	12	17.716	8	3127.92	6	2231.34	2
27	14.732	6	17.723	7	3470.8	4	2485.15	5
28	15.415	10	16.008	2	4123.21	2	2202.26	4
29	16.072	11	20.001	3	3045.36	3	1906.6	3
30	17.707	10	15.817	7	3073.86	11	1894.66	2
31	17.289	10	12.271	2	2226.16	2	1707.57	2
Mean	9.26		7		5.03		3.65	
STD	3.22		1.88		3.32		1.54	

Table 4.20 Results of FFWAC-III for wine dataset using Inter-cluster

Distance	Minimum				Maximum			
Metric	$\sqrt{n}$		*Sturges*		$\sqrt{n}$		*Sturges*	
Fitness/ number	Fit	K	Fit	K	Fit	K	Fit	K
1	7.33	2	11.773	4	3490.58	3	2597.91	8
2	7.45	5	6.888	7	3634.41	2	1740.96	2
3	31.16	6	0.380	2	3524.12	3	2166.01	2
4	0.40	2	12.486	3	3241.75	4	2115.02	2
5	2.86	4	26.389	4	3285.33	5	1839.02	2
6	20.53	3	0.671	2	3502.84	6	2038.63	2
7	31.45	10	21.137	4	3688.94	13	2330.08	6
8	9.94	6	7.666	5	3814.00	9	2170.09	5
9	0.18	2	8.790	4	2673.20	8	2347.23	2
10	7.62	9	4.556	4	3147.21	7	2279.2	2
11	15.30	3	4.160	2	3026.00	3	2351.02	2
12	17.93	5	0.483	2	3900.29	5	2137.8	3
13	44.60	11	3.058	3	2777.14	3	2130.18	5
14	5.59	8	0.165	3	2520.69	4	1979.99	7
15	2.45	4	6.430	3	3768.29	6	2560.51	3
16	36.63	7	0.005	2	3584.82	3	2265.16	5
17	3.71	4	1.690	2	3707.92	2	2383.69	3
18	5.25	5	0.905	2	3480.58	7	1836.24	4
19	29.66	10	0.595	2	3344.86	9	2593.17	3
20	10.71	3	17.622	3	3710.71	2	1501.09	4
21	28.08	3	1.835	2	3664.64	2	1732.42	3
22	0.60	2	0.584	2	2403.96	2	2402.69	2
23	37.93	7	30.491	5	3573.31	3	2082.87	3
24	21.40	9	1.650	8	2697.06	2	2639.48	3
25	2.17	2	0.035	2	2346.32	2	1513.67	2
26	26.42	3	21.406	3	3378.02	4	2476.27	7
27	42.52	6	0.928	5	3409.81	4	2161.53	7
28	26.00	6	10.021	3	3223.43	2	1946.98	2
29	45.74	9	9.944	4	2555.23	4	2160.57	2
30	17.89	3	25.582	5	3163.50	6	2462.81	7
31	4.00	2	6.013	4	3102.99	8	2166.41	3
Mean	5.19		**3.42**		4.61		**3.65**	
STD	2.80		**1.52**		2.72		**1.92**	

Table 4.21 Results of FFWAC-I for WDBC dataset using Intra-cluster

Distance	Minimum				Maximum			
Metric	$\sqrt{n}$		*Sturges*		$\sqrt{n}$		*Sturges*	
Fitness/number	Fit	K	Fit	K	Fit	K	Fit	K
1	166.99	14	47.74	2	2064.8	15	2295.64	7
2	124.99	13	196.46	9	1103.5	9	2104.57	9
3	182.48	12	31.38	4	1847.4	19	1651.94	9
4	149.02	20	37.15	8	3234.6	15	1605.72	9
5	134.15	20	40.06	2	1806.9	15	3685.80	9
6	147.58	18	37.37	2	1066.5	14	1503.54	9
7	146.45	17	62.96	9	1249.3	20	2179.92	5
8	105.98	3	199.81	8	1747.3	22	1753.13	3
9	169.41	17	129.36	9	990.2	23	2581.49	10
10	130.87	15	31.99	3	1089.5	19	1747.94	3
11	121.36	13	197.73	5	1607.6	23	1593.81	5
12	142.96	3	23.81	10	1446.1	15	1434.09	8
13	164.95	18	28.29	10	1425.7	15	2431.98	8
14	140.19	20	27.91	5	1574.8	23	2282.61	7
15	151.17	18	161.31	10	949.6	18	1476.84	7
16	145.59	22	41.65	6	3988.0	22	1522.59	7
17	44.40	17	84.92	3	1781.8	20	3223.49	9
18	103.86	19	31.00	7	1399.0	14	1980.21	9
19	27.94	20	141.66	8	2112.5	11	1620.99	2
20	161.06	14	189.33	9	2509.0	8	1309.57	7
21	171.64	13	34.61	6	2128.5	3	1518.46	7
22	24.57	6	200.96	7	1805.4	5	1903.64	8
23	147.94	22	23.80	2	1700.7	3	1870.04	8
24	167.18	15	33.17	5	1429.9	7	2269.47	7
25	148.29	22	176.46	7	1114.8	23	1513.57	7
26	178.52	9	152.03	6	1842.5	24	1435.20	7
27	140.24	17	24.49	5	1722.4	2	1777.03	9
28	174.53	18	176.64	5	1274.1	12	1381.18	8
29	158.16	15	26.41	9	1644.4	17	4022.11	9
30	137.70	16	107.32	6	1255.8	14	2636.61	4
31	134.92	8	46.72	3	1056.7	19	2947.67	5
Mean	15.29		6.11		15.1		7.13	
STD	5.13		2.51		6.5		2.06	

Table 4.22 Results of FFWAC-I for WDBC dataset using Inter-cluster

Distance	Minimum				Maximum			
Metric	$\sqrt{n}$		*Sturges*		$\sqrt{n}$		*Sturges*	
Fitness/ number	Fit	K	Fit	K	Fit	K	Fit	K
1	47.420	11	22.729	9	27,071.11	7	9114.684	4
2	166.361	21	29.225	4	16,460.23	3	13,117.992	4
3	32.166	7	155.201	10	28,975.14	4	10,411.008	5
4	144.455	9	26.803	4	30,320.72	2	11,162.007	2
5	162.154	12	30.984	6	26,071.14	4	7217.242	3
6	31.202	15	24.101	4	27,488.38	5	12,692.544	8
7	129.127	21	40.151	5	28,239.24	6	13,193.462	4
8	43.900	2	161.001	9	32,456.98	24	10,694.813	2
9	27.191	18	25.820	5	24,245.43	5	8373.473	3
10	134.438	23	154.693	9	28,862.22	8	13,531.131	6
11	145.383	21	43.186	3	28,628.58	5	11,833.871	2
12	22.540	3	157.522	8	28,125.31	8	11,409.729	2
13	27.641	9	45.981	6	27,172.91	15	7917.673	2
14	185.525	3	36.748	3	28,260.13	8	13,487.203	9
15	30.404	18	173.074	6	25,354.71	18	11,026.349	4
16	167.429	16	164.509	10	27,827.28	2	9956.647	8
17	39.593	10	34.025	6	24,028.88	2	7532.554	5
18	137.342	15	23.628	6	26,494.20	18	13,769.499	5
19	146.083	21	22.073	8	25,946.40	5	7490.365	3
20	139.191	23	201.558	9	28,372.86	24	11,117.693	2
21	160.851	15	23.475	4	26,948.81	18	11,690.336	3
22	106.104	22	35.019	5	27,961.21	10	9821.677	7
23	149.586	18	25.957	4	26,525.58	12	12,523.558	3
24	143.424	16	28.741	3	33,508.60	24	11,181.053	2
25	153.057	20	28.152	4	21,949.66	2	9699.145	2
26	139.925	19	20.690	9	25,707.94	8	14,673.479	2
27	139.526	18	32.751	2	26,646.72	4	7791.942	3
28	37.473	7	147.933	8	26,183.57	3	12,928.639	2
29	125.018	17	28.445	3	31,652.23	8	10,391.139	4
30	149.412	20	191.117	9	27,158.27	7	12,978.711	8
31	47.241	2	152.141	9	21,428.04	6	14,193.305	10
Mean	14.58		6.13		8.87		4.16	
STD	6.58		2.49		6.83		2.38	

Table 4.23 Results of FFWAC-II for WDBC dataset using Intra-cluster

Distance	Minimum				Maximum			
Metric	$\sqrt{n}$		*Sturges*		$\sqrt{n}$		*Sturges*	
Fitness/number	Fit	K	Fit	K	Fit	K	Fit	K
1	166.99	14	47.74	2	1366.53	18	1583.37	7
2	124.99	13	196.46	9	1526.30	12	1412.38	7
3	182.48	12	31.38	4	1741.72	22	1665.83	9
4	149.02	20	37.15	8	1078.81	7	2156.02	2
5	134.15	20	40.06	2	2051.59	15	1335.76	8
6	147.58	18	37.37	2	1302.25	21	1858.59	7
7	146.45	17	62.96	9	1157.75	20	2244.67	6
8	105.98	3	199.81	8	1631.65	23	2517.60	6
9	169.41	17	129.36	9	2502.03	17	3111.14	7
10	130.87	15	31.99	3	993.42	19	3707.71	5
11	121.36	13	197.73	5	1215.55	17	3871.41	10
12	142.96	3	23.81	10	1727.76	23	1618.43	7
13	164.95	18	28.29	10	2291.47	14	2743.12	3
14	140.19	20	27.91	5	1424.34	20	3530.70	10
15	151.17	18	161.31	10	2153.06	22	1411.43	9
16	145.59	22	41.65	6	1114.61	12	1577.33	7
17	44.40	17	84.92	3	2234.45	23	1732.44	8
18	103.86	19	31.00	7	1103.88	9	2462.71	9
19	27.94	20	141.66	8	1127.50	18	1432.78	5
20	161.06	14	189.33	9	1158.13	19	2199.71	8
21	171.64	13	34.61	6	1425.41	21	4190.21	4
22	24.57	6	200.96	7	1272.00	15	1771.61	10
23	147.94	22	23.80	2	1510.47	23	3178.54	10
24	167.18	15	33.17	5	1287.69	20	1555.71	7
25	148.29	22	176.46	7	2949.74	23	1746.53	7
26	178.52	9	152.03	6	934.61	13	2144.90	7
27	140.24	17	24.49	5	2459.39	21	2767.03	9
28	174.53	18	176.64	5	2323.62	21	2103.82	10
29	158.16	15	26.41	9	1339.02	20	2705.64	9
30	137.70	16	107.32	6	1103.42	13	2203.60	8
31	134.92	8	46.72	3	1107.63	11	1570.37	10
Mean	14.28		6.13		17.81		7.45	
STD	5.48		2.63		4.54		2.10	

Table 4.24 Results of FFWAC-II for WDBC dataset using Inter-cluster

Distance	Minimum				Maximum			
Metric	$\sqrt{n}$		*Sturges*		$\sqrt{n}$		*Sturges*	
Fitness/number	Fit	K	Fit	K	Fit	K	Fit	K
1	110.62	11	22.342	10	27,229.7	10	8466.58694	5
2	155.99	17	28.339	6	28,855.6	7	17,645.3055	3
3	31.21	15	27.102	7	29,180.4	3	9417.11741	2
4	168.68	16	170.740	7	27,330.9	20	7184.43	2
5	39.38	20	36.917	5	26,271.5	11	13,977.68	3
6	157.24	19	24.879	5	27,668.5	6	11,326.44	3
7	151.75	20	177.780	4	26,421.7	2	11,489.64	3
8	155.63	18	44.056	4	24,601.1	4	13,214.60	4
9	167.84	14	46.506	2	31,391.3	3	12,299.61	3
10	148.30	23	199.386	10	25,373.5	12	9341.53	2
11	154.00	13	25.696	5	25,014.1	8	12,382.66	3
12	157.95	10	35.139	5	33,228.7	2	10,011.10	7
13	142.73	14	183.216	4	25,970.6	3	11,726.12	6
14	44.04	4	23.596	7	32,984.2	8	12,281.69	4
15	146.16	22	172.321	7	28,953.1	22	11,978.28	2
16	75.02	2	25.984	8	29,665.5	23	12,112.20	7
17	144.31	13	153.383	9	29,503.8	2	13,589.23	8
18	37.99	9	55.946	2	26,636.5	2	14,133.22	6
19	125.33	18	40.643	3	29,032.4	10	8297.95	5
20	34.76	5	24.718	7	23,079.4	4	11,741.19	8
21	128.59	21	29.478	2	22,548.4	17	9232.28	3
22	141.30	18	25.350	6	30,881.2	9	11,800.67	2
23	27.12	6	29.761	3	26,197.8	3	11,087.35	3
24	30.22	3	103.974	2	29,947.6	20	10,930.48	5
25	137.92	23	23.517	2	21,213.6	2	11,272.47	2
26	33.87	5	49.521	3	26,437.3	21	10,427.14	5
27	27.97	3	36.306	4	25,724.5	4	10,898.12	3
28	145.67	19	31.455	6	29,172.8	5	9784.59	2
29	22.30	5	33.668	9	31,230.7	24	12,335.67	2
30	36.64	10	81.476	3	29,670.7	6	9563.90	3
31	27.46	7	135.345	9	30,848.9	17	12,030.31	3
Mean	13.01		5.35		9.35		**3.84**	
STD	6.68		5.35		7.35		**1.86**	

Table 4.25 Results of FFWAC-III for WDBC dataset using Intra-cluster

Distance	Minimum				Maximum			
Metric	$\sqrt{n}$		*Sturges*		$\sqrt{n}$		*Sturges*	
Fitness/number	Fit	K	Fit	K	Fit	K	Fit	K
1	148.30	23	29.860	3	28,335.82	2	11,563.72	5
2	154.00	13	13.350	4	31,187.56	15	10,267.17	3
3	157.95	10	58.900	6	31,097.56	12	14,887.92	2
4	142.73	14	18.310	8	26,956.16	10	11,292.94	8
5	44.04	4	56.180	4	28,362.27	9	11,870.89	3
6	146.16	22	16.210	6	23,395.06	7	12,924.32	3
7	75.02	2	2.100	2	28,541.41	9	12,687.92	9
8	144.31	13	8.730	2	25,060.71	5	9048.53	4
9	37.99	9	75.600	3	29,770.16	10	7096.35	4
10	125.33	18	3.630	3	27,308.43	8	13,534.93	9
11	34.76	5	33.690	6	29,682.18	17	12,509.79	2
12	128.59	21	0.530	2	24,134.95	8	11,345.19	3
13	141.30	18	0.510	2	33,754.66	20	13,203.05	3
14	27.12	6	1.730	3	24,852.04	19	12,265.42	4
15	30.22	3	32.570	4	29,710.60	3	10,046.62	2
16	137.92	23	2.280	2	19,094.30	15	9113.49	6
17	33.87	5	3.590	6	27,820.37	19	13,068.43	3
18	27.97	3	23.670	6	25,507.82	10	9910.86	3
19	145.67	19	23.830	3	31,420.37	9	8847.79	4
20	22.30	5	2.300	2	29,723.73	14	10,317.40	6
21	110.62	11	23.400	11	25,152.42	9	12,775.73	9
22	155.99	17	15.350	10	24,177.24	2	11,264.79	4
23	31.21	15	18.580	4	30,152.13	8	13,445.44	2
24	168.68	16	5.440	4	17,859.53	10	8973.33	7
25	39.38	20	4.890	4	26,691.29	3	13,716.15	4
26	157.24	19	3.920	3	26,760.57	9	12,628.42	2
27	151.75	20	34.040	9	19,357.98	2	12,140.11	3
28	155.63	18	0.000	2	24,760.31	20	10,345.95	6
29	167.84	14	6.910	8	28,230.09	17	9448.32	3
30	148.30	23	28.750	3	26,951.67	10	13,446.74	3
31	110.62	11	6.790	5	23,039.35	19	8203.03	5
Mean	12.26		4.77		10.65		4.32	
STD	6.68		4.89		5.59		2.17	

Table 4.26 Results of FFWAC-III for WDBC dataset using Inter-cluster

Distance	Minimum				Maximum			
Metric	$\sqrt{n}$		*Sturges*		$\sqrt{n}$		*Sturges*	
Fitness/ number	Fit	K	Fit	K	Fit	K	Fit	K
1	72.69	9	9.63	3	25,898.64	4	12,606.29	2
2	53.77	3	0.33	2	30,704.54	14	7325.55	2
3	12.35	3	1.85	2	30,248.95	3	10,339.27	2
4	10.00	2	14.42	4	27,441.39	9	11,577.77	3
5	125.80	5	77.48	5	33,792.52	5	7286.99	4
6	92.42	18	14.05	3	29,185.09	5	8986.97	2
7	539.61	18	6.58	3	29,607.32	7	12,966.25	2
8	158.77	5	0.80	2	27,940.40	3	7639.66	3
9	423.88	14	589.58	8	29,278.17	7	9745.94	5
10	88.20	3	483.26	6	30,420.81	16	12,933.66	9
11	293.32	6	68.29	3	34,798.48	19	6119.99	3
12	322.14	3	11.56	2	31,015.76	7	9763.12	4
13	42.53	3	16.51	2	25,149.28	4	14,873.95	4
14	185.04	7	120.89	6	29,905.21	22	9174.81	2
15	220.51	10	25.29	7	33,712.65	11	9227.83	5
16	73.28	8	91.10	3	18,070.80	3	8361.58	6
17	357.15	8	173.21	5	29,945.34	12	11,644.28	8
18	147.28	5	37.58	3	22,077.56	6	12,581.07	2
19	338.59	11	247.78	9	28,133.17	4	11,436.23	6
20	76.19	3	22.51	2	28,018.38	4	12,489.96	8
21	231.57	4	47.34	7	25,999.96	3	10,805.01	8
22	928.86	9	2.66	2	27,133.16	18	9507.54	3
23	2.48	2	122.09	5	32,172.94	9	12,826.44	2
24	266.27	12	0.06	2	22,455.87	9	12,675.25	2
25	583.71	7	32.10	4	29,622.32	22	11,109.46	5
26	449.74	13	11.86	2	23,458.30	5	13,264.60	2
27	147.68	5	200.50	5	27,216.58	4	13,728.97	4
28	57.98	6	177.42	10	22,067.72	19	10,576.67	3
29	39.77	13	35.34	4	30,956.56	19	13,582.84	4
30	214.07	5	31.18	3	25,221.17	9	12,089.39	2
31	107.02	4	271.56	4	26,382.60	2	13,687.76	3
Mean	7.23		4.13		9.16		**3.87**	
STD	4.50		2.23		6.30		**2.11**	

As the same of FFWAC-II, the best result obtained of the FFWAC-III performance for WDBC dataset is using Inter-cluster with maximum distance in Sturges Law.

4.1.3 Interval Type 2 Fuzzy Logic Fireworks Algorithm for Clustering (F2FFWA)

The third variation of FWA is the Interval Type 2 Fuzzy Logic for Clustering. In Table 4.27 their parameters are shown.

The results obtained of F2FWAC for Iris dataset using Intra-cluster and Inter-cluster are shown in Tables 4.28 and 4.29 with FFWAC-I, Tables 4.30 and 4.31 with FFWAC-II, and, Tables 4.32 and 4.33 (FFWAC-III), respectively.

Tables 4.15, 4.17 and 4.19 show the results using Intra-cluster; Tables 4.16, 4.18 and 4.20 using Inter-cluster of the FFWAC-I, FFWAC-II and FFWAC-III, respectively. The results are obtained for the wine dataset.

For WDBC dataset, Tables 4.21, 4.23 and 4.25 show the result using Intra-cluster of FFWAC-I, FFWAC-II and FFWAC-III, respectively; and, Tables 4.22, 4.24 and 4.26 show results using Inter-cluster of FFWAC-I, FFWAC-II and FFWAC-III, respectively.

4.1.3.1 Iris Dataset

F2FFWAC-I obtained the best result for the Iris dataset using Inter-cluster and the Sturges Law with maximum distance, as marked in Table 4.29 with bold type.

As the same of F2FWAC-I, F2FWAC-II, obtained the best result for the Iris dataset using Inter-cluster and the Sturges Law with maximum distance.

Table 4.33 shows the best result obtained of F2FWAC-II for the Iris dataset using Inter-cluster and the Sturges Law with maximum distance (Table 4.34)

Table 4.27 Parameters of F2FWAC

Parameter	Value
Fireworks (X_i)	5
Amplitude coefficient ($\widehat{A}$)	[40, 2]
Spark coefficient (m)	50
Dimensions (d)	Features data
Function evaluations	15,000
Membership functions (interval type 2 fuzzy logic)	• Triangular • Gaussian • Trapezoidal

Table 4.28 Results of F2FWAC-I for Iris dataset using Intra-cluster

Distance	Minimum				Maximum			
Metric	$\sqrt{n}$		*Sturges*		$\sqrt{n}$		*Sturges*	
Fitness/number	Fit	K	Fit	K	Fit	K	Fit	K
1	0.596	8	0.526	7	4.98	4	4.84	8
2	0.432	2	0.475	3	4.73	9	4.74	2
3	0.446	3	0.503	5	4.76	10	5.09	6
4	0.456	6	0.432	4	5.33	2	4.95	8
5	0.452	6	0.548	3	4.88	3	4.72	7
6	0.590	10	0.550	5	4.74	7	4.73	6
7	0.490	2	0.570	6	5.05	2	4.91	3
8	0.417	2	0.510	3	4.92	11	4.88	4
9	0.553	3	0.557	7	4.80	5	4.81	2
10	0.572	11	0.560	8	4.85	7	4.96	6
11	0.514	8	0.540	7	4.82	8	4.46	6
12	0.610	11	0.476	5	4.86	5	4.59	5
13	0.585	10	0.545	3	4.81	11	4.98	5
14	0.536	10	0.586	7	5.24	3	4.69	6
15	0.487	10	0.605	7	4.76	3	4.77	5
16	0.461	3	0.371	4	5.57	10	4.73	4
17	0.528	12	0.463	2	4.58	11	4.63	8
18	0.549	6	0.594	6	4.63	3	4.76	3
19	0.359	2	0.500	3	4.33	9	4.71	2
20	0.548	7	0.519	2	4.82	4	4.77	6
21	0.516	3	0.498	4	4.66	8	4.52	4
22	0.501	3	0.492	5	5.07	9	4.79	5
23	0.493	11	0.498	2	4.77	2	4.81	5
24	0.526	9	0.456	2	5.40	5	5.01	4
25	0.481	5	0.464	4	4.75	9	5.02	6
26	0.441	5	0.510	4	4.82	4	4.74	4
27	0.537	11	0.510	8	4.55	7	5.36	5
28	0.468	7	0.430	3	4.33	10	4.96	6
29	0.566	2	0.490	3	5.06	10	4.75	2
30	0.188	3	0.570	4	4.90	9	4.64	8
31	0.636	6	0.480	3	4.56	8	4.92	5
Mean	6.35		4.48		6.71		5.03	
STD	3.42		1.88		3.07		1.80	

Table 4.29 Results of F2FWAC-I for Iris dataset using Inter-cluster

Distance	Minimum				Maximum			
Metric	$\sqrt{n}$		Sturges		$\sqrt{n}$		Sturges	
Fitness/number	Fit	K	Fit	K	Fit	K	Fit	K
1	0.432	11	0.590	3	29.304	4	16.359	3
2	0.516	2	0.536	6	36.637	3	22.319	3
3	0.588	9	0.524	4	26.611	3	17.321	2
4	0.593	10	0.601	4	26.266	4	16.383	2
5	0.623	9	0.575	7	27.978	4	21.701	2
6	0.514	7	0.180	2	20.338	6	20.824	3
7	0.513	10	0.541	4	16.777	3	14.690	2
8	0.553	10	0.493	5	24.389	9	17.396	2
9	0.384	3	0.526	7	27.495	4	14.970	2
10	0.308	2	0.534	3	27.739	4	13.175	3
11	0.521	10	0.496	4	25.000	6	16.185	4
12	0.290	2	0.523	8	27.283	4	18.153	6
13	0.628	6	0.555	4	29.036	4	21.260	6
14	0.588	7	0.458	7	23.036	4	15.246	5
15	0.545	11	0.578	4	29.172	2	15.347	3
16	0.580	5	0.593	2	21.935	5	16.277	3
17	0.528	10	0.436	2	23.427	4	19.289	5
18	0.600	12	0.428	2	25.019	4	17.042	3
19	0.488	9	0.566	8	19.254	4	12.373	5
20	0.571	12	0.489	5	24.387	3	22.931	5
21	0.538	10	0.568	5	19.568	4	14.463	3
22	0.469	4	0.535	4	29.090	4	20.240	2
23	0.574	10	0.567	4	26.789	6	14.976	3
24	0.509	5	0.371	2	23.381	3	19.222	3
25	0.539	9	0.490	7	22.057	2	19.711	6
26	0.470	7	0.504	5	22.315	2	18.206	2
27	0.319	3	0.500	6	29.706	3	12.149	2
28	0.532	10	0.596	8	24.950	3	17.971	2
29	0.142	2	0.616	6	17.646	3	18.799	4
30	0.610	9	0.336	2	22.682	10	21.367	6
31	0.563	9	0.552	4	24.961	3	16.049	4
Mean	7.58		4.65		4.10		**3.42**	
STD	3.25		1.94		1.78		**1.41**	

Table 4.30 Results of F2FWAC-II for Iris dataset using Intra-cluster

Distance	Minimum				Maximum			
Metric	$\sqrt{n}$		*Sturges*		$\sqrt{n}$		*Sturges*	
Fitness/number	Fit	K	Fit	K	Fit	K	Fit	K
1	0.5	6	0.571	7	4.94	3	4.81	3
2	0.54	3	0.559	7	4.68	11	4.71	3
3	0.49	9	0.549	5	4.74	10	4.88	4
4	0.56	10	0.470	4	4.98	4	4.82	7
5	0.57	6	0.367	2	4.82	6	4.94	3
6	0.56	11	0.525	3	4.42	8	4.66	6
7	0.46	11	0.363	5	4.50	9	4.82	8
8	0.6	6	0.376	3	4.66	6	4.79	3
9	0.6	11	0.574	3	4.49	7	4.53	7
10	0.48	4	0.543	2	4.75	9	4.7	5
11	0.55	12	0.596	5	4.48	11	4.88	4
12	0.53	11	0.536	6	4.74	4	4.68	2
13	0.49	2	0.601	7	4.66	3	4.8	6
14	0.47	6	0.428	4	4.69	5	4.81	8
15	0.63	4	0.560	8	4.78	12	4.68	3
16	0.53	3	0.585	4	4.74	6	4.51	8
17	0.39	3	0.523	5	4.59	7	4.86	2
18	0.57	2	0.559	6	4.86	8	4.74	7
19	0.54	11	0.567	7	4.83	3	4.54	5
20	0.59	11	0.518	4	4.78	8	5.49	5
21	0.49	9	0.492	3	5.05	3	5.18	4
22	0.5	7	0.541	5	4.01	6	4.79	7
23	0.53	4	0.534	6	4.89	5	4.81	7
24	0.56	6	0.606	4	5.43	9	4.86	2
25	0.56	11	0.515	5	4.75	6	4.77	5
26	0.47	9	0.524	7	3.96	11	4.81	2
27	0.47	9	0.466	8	4.76	7	4.81	7
28	0.57	5	0.440	3	4.76	12	4.86	3
29	0.54	12	0.460	4	4.65	2	4.63	6
30	0.55	11	0.532	7	4.65	8	5.43	6
31	0.55	8	0.475	6	4.88	11	4.52	4
Mean	7.52		5		7.10		4.9	
STD	3.32		1.73		2.94		1.99	

Table 4.31 Results of F2FWAC-II for Iris dataset using Inter-cluster

Distance	Minimum				Maximum			
Metric	$\sqrt{n}$		Sturges		$\sqrt{n}$		Sturges	
Fitness/number	Fit	K	Fit	K	Fit	K	Fit	K
1	0.430	3	0.404	2	25.258	4	16.544	3
2	0.457	3	0.544	8	31.003	2	16.683	2
3	0.523	5	0.570	4	23.000	4	15.861	2
4	0.519	7	0.592	7	24.069	3	18.249	4
5	0.658	10	0.615	5	28.895	3	18.373	3
6	0.479	6	0.568	3	26.467	7	18.882	3
7	0.470	3	0.557	3	29.815	5	14.873	2
8	0.536	7	0.529	2	31.263	3	22.369	3
9	0.573	10	0.498	7	25.613	10	15.703	3
10	0.606	8	0.586	7	25.176	3	16.431	2
11	0.587	2	0.568	8	23.959	5	18.347	4
12	0.581	10	0.510	5	20.629	2	20.570	7
13	0.496	11	0.501	8	23.183	5	13.549	5
14	0.620	6	0.548	7	26.416	2	21.492	3
15	0.525	9	0.508	5	30.009	3	18.322	2
16	0.590	12	0.563	5	25.772	6	21.957	4
17	0.521	4	0.522	2	16.938	3	20.872	5
18	0.517	4	0.389	2	26.275	6	10.703	2
19	0.051	2	0.542	4	18.000	5	17.071	5
20	0.513	5	0.497	4	25.439	7	13.660	4
21	0.573	5	0.528	2	27.679	2	17.058	3
22	0.575	6	0.494	5	26.999	8	9.773	2
23	0.562	8	0.522	7	16.388	2	17.361	3
24	0.425	3	0.432	2	18.453	5	15.950	2
25	0.590	9	0.444	4	28.046	2	16.214	3
26	0.238	2	0.505	3	29.340	4	14.440	4
27	0.278	3	0.583	5	27.113	5	17.430	3
28	0.555	5	0.521	4	18.039	2	16.891	3
29	0.516	7	0.504	4	18.357	3	18.676	2
30	0.507	2	0.581	4	30.502	9	18.625	3
31	0.586	12	0.424	2	28.732	3	14.623	3
Mean	6.10		4.52		4.29		**3.19**	
STD	3.13		2.01		2.18		**1.17**	

Table 4.32 Results of F2FWAC-III for Iris dataset using Intra-cluster

Distance	Minimum				Maximum			
Metric	$\sqrt{n}$		Sturges		$\sqrt{n}$		Sturges	
Fitness/number	Fit	K	Fit	K	Fit	K	Fit	K
1	0.45	3	0.58	3	4.88	9	5.06	5
2	0.50	4	0.57	7	4.80	4	5.49	7
3	0.55	8	0.52	5	4.75	4	4.71	3
4	0.52	8	0.50	5	4.50	11	4.48	4
5	0.53	3	0.65	5	4.57	11	4.99	4
6	0.39	3	0.56	8	5.39	2	4.93	6
7	0.50	10	0.47	3	4.64	4	4.66	4
8	0.55	5	0.57	7	4.75	12	4.91	7
9	0.61	9	0.63	8	3.53	10	5.02	5
10	0.58	11	0.58	7	4.79	4	4.88	2
11	0.59	5	0.42	2	4.68	8	4.52	8
12	0.60	2	0.49	2	4.80	7	5.05	6
13	0.53	10	0.53	3	4.62	5	4.79	4
14	0.43	4	0.41	3	3.28	11	3.98	7
15	0.44	4	0.60	3	4.75	7	4.95	3
16	0.51	9	0.56	4	4.81	7	4.70	6
17	0.48	3	0.41	3	4.89	4	4.69	3
18	0.49	3	0.46	4	4.82	9	4.71	7
19	0.47	5	0.63	5	4.81	6	4.93	5
20	0.61	7	0.58	7	4.36	11	4.66	4
21	0.53	7	0.51	3	4.60	5	4.76	6
22	0.59	9	0.60	6	4.86	10	4.83	7
23	0.57	8	0.50	4	4.78	11	4.91	5
24	0.55	7	0.48	2	4.74	3	4.56	3
25	0.56	8	0.57	3	4.61	10	5.06	7
26	0.56	3	0.46	5	4.61	3	5.16	3
27	0.53	10	0.43	2	4.55	4	4.61	5
28	0.45	5	0.58	4	4.79	12	4.82	3
29	0.49	9	0.61	4	4.77	10	5.27	7
30	0.57	10	0.48	6	4.72	3	4.82	6
31	0.53	10	0.39	2	4.58	7	4.58	4
Mean	6.52		4.35		7.23		5.03	
STD	2.82		1.87		3.21		1.64	

Table 4.33 Results of F2FWAC-III for Iris dataset using Inter-cluster

Distance	Minimum				Maximum			
Metric	$\sqrt{n}$		*Sturges*		$\sqrt{n}$		*Sturges*	
Fitness/number	Fit	K	Fit	K	Fit	K	Fit	K
1	0.498	6	0.477	3	21.145	4	15.880	3
2	0.509	10	0.506	3	30.144	3	18.220	3
3	0.371	2	0.556	6	17.121	5	14.540	3
4	0.541	2	0.428	2	22.983	5	10.148	3
5	0.588	9	0.500	2	26.978	2	16.276	3
6	0.482	2	0.054	2	28.120	4	18.932	4
7	0.445	3	0.113	2	25.249	2	17.957	3
8	0.541	10	0.552	8	19.311	4	17.755	2
9	0.559	7	0.531	4	14.519	3	21.284	2
10	0.528	3	0.564	5	25.133	4	17.132	2
11	0.290	2	0.479	5	26.010	2	19.706	2
12	0.384	2	0.594	3	28.075	3	13.750	3
13	0.464	6	0.518	4	28.705	2	17.665	3
14	0.525	7	0.535	6	23.041	4	22.654	2
15	0.594	8	0.449	3	26.886	4	16.580	2
16	0.573	9	0.568	4	26.703	5	15.413	2
17	0.518	7	0.538	3	31.291	4	17.969	2
18	0.547	6	0.485	4	28.297	10	17.400	2
19	0.517	5	0.486	3	19.404	3	20.586	4
20	0.514	11	0.456	3	24.467	4	16.017	2
21	0.553	3	0.546	5	30.361	3	21.513	6
22	0.433	2	0.533	2	24.467	5	17.442	2
23	0.401	4	0.514	2	27.375	2	17.947	3
24	0.566	7	0.610	7	25.734	3	17.376	3
25	0.515	8	0.559	8	29.737	3	18.884	2
26	0.609	2	0.563	7	30.605	3	15.955	4
27	0.574	4	0.524	3	28.905	11	22.612	4
28	0.495	2	0.510	2	30.504	5	14.889	3
29	0.589	4	0.534	5	28.653	3	15.487	3
30	0.613	10	0.539	7	27.545	11	11.358	3
31	0.519	6	0.040	2	22.902	5	20.144	5
Mean	5.45		4.03		4.23		**2.90**	
STD	2.95		1.92		2.36		**0.98**	

Table 4.34 Results of F2FWAC-I for wine dataset using Intra-cluster

Distance	Minimum				Maximum			
Metric	$\sqrt{n}$		*Sturges*		$\sqrt{n}$		*Sturges*	
Fitness/number	Fit	K	Fit	K	Fit	K	Fit	K
1	19.00	6	20.01	8	238.01	11	4.81	3
2	16.80	12	16.87	4	309.91	9	4.71	3
3	16.15	7	18.66	6	474.23	13	4.88	4
4	14.83	12	13.22	3	251.76	4	4.82	7
5	17.71	12	19.45	6	261.83	12	4.94	3
6	17.08	8	19.36	6	294.45	10	4.66	6
7	16.14	13	19.39	5	319.59	11	4.82	8
8	15.86	12	15.14	7	560.95	6	4.79	3
9	19.66	9	17.51	4	215.72	10	4.53	7
10	17.36	12	18.08	7	200.29	10	4.70	5
11	16.94	9	10.92	5	228.61	11	4.88	4
12	16.57	12	18.60	7	521.95	7	4.68	2
13	13.80	11	20.89	7	308.23	10	4.80	6
14	16.84	10	16.85	4	516.90	12	4.81	8
15	17.03	10	18.67	6	403.53	11	4.68	3
16	16.07	5	20.52	6	301.38	11	4.51	8
17	18.50	7	16.56	8	253.43	4	4.86	2
18	12.70	3	17.06	4	345.05	5	4.74	7
19	18.01	11	15.38	8	398.89	10	4.54	5
20	15.88	6	24.88	3	212.57	13	5.49	5
21	15.22	11	21.05	3	293.21	11	5.18	4
22	17.98	5	17.52	8	243.34	9	4.79	7
23	19.44	11	22.01	5	241.23	13	4.81	7
24	16.91	4	18.99	3	231.19	11	4.86	2
25	17.52	9	17.86	7	413.77	11	4.77	5
26	16.52	8	19.25	7	295.44	10	4.81	2
27	16.59	9	22.21	6	268.80	7	4.81	7
28	15.22	11	19.54	4	449.19	9	4.86	3
29	20.91	6	19.90	5	301.70	4	4.63	6
30	15.29	12	15.77	8	494.05	11	5.43	6
31	18.18	12	17.22	8	278.35	3	4.52	4
Mean	6.38		5.74		9.32		4.90	
STD	3.1		1.71		2.88		1.99	

4.1.3.2 Wine Dataset

For the wine dataset, the best result is shown in Table 4.35. FWAC-I obtained the best performance using Inter-cluster and the Sturges Law with maximum distance (Table 4.36).

For the wine dataset, the best results of the F2FWAC-II and F2FC-III are shown in Tables 4.38 and 4.39, respectively. For both algorithms, the best performance is using Inter-cluster and the Sturges Law with maximum distance; the results are marked in bold type (Table 4.37).

4.1.3.3 Breast Cancer Wisconsin Diagnostic (WDBC)

Table 4.40 shows the results of F2FWAC-I for WDBC.

For the WDBC dataset, most of the results were not satisfactory, but the best result obtained by the F2FWAC-I, is using Inter-cluster with the Sturges Law and maximum distance, as is shown in Table 4.41. The results of F2FWAC-II for WDBC are presented in Table 4.42.

Table 4.43 shows the best result obtained of the F2FWAC-II for the WDBC dataset, which is to use Inter-cluster with Sturges Law and maximum distance.

The results of the F2FWAC-III shown in Tables 4.44 and 4.45 were not satisfactory for the WDBC dataset.

4.2 Classification Methods

In this section we show the results obtained with the Type 1 Fuzzy Classifiers (T1FIS) and Interval Type 2 Fuzzy Classifiers (IT2FIS), both Mamdani and Sugeno type.

4.2.1 Type 1 FIS Classifier (T1FIS)

The best results of the T1FIS for the Iris dataset are shown in Table 4.46, both, Mamdani and Sugeno type are obtained using Gaussian MFs; the results are marked in bold type. Table 4.47 shows the best results with Gaussian MFs in T1FIS for the wine dataset.

Table 4.48 shows the T1FIS results for the WDBC dataset. According with the results we can notice that Gaussian MFs in Mamdani type are the best with a mean classification of 92.33%, and, Trapezoidal MFs in Sugeno type are the best with a mean classification of 93.91%.

Table 4.35 Results of F2FWAC-I for wine dataset using Inter-cluster

Distance	Minimum				Maximum			
Metric	$\sqrt{n}$		*Sturges*		$\sqrt{n}$		*Sturges*	
Fitness/ number	Fit	K	Fit	K	Fit	K	Fit	K
1	21.744	12	17.923	8	3945.88	3	2023.086	4
2	18.452	10	21.448	3	3685.54	3	1920.022	4
3	16.154	6	16.705	5	3929.66	7	2013.142	2
4	15.114	11	17.022	4	2727.94	2	2441.549	2
5	15.114	12	0.651	2	3194.02	4	1770.795	3
6	14.361	12	18.670	7	3474.01	3	2152.688	4
7	18.991	8	17.463	7	3475.22	7	2353.596	2
8	16.967	8	20.156	6	1992.18	2	2209.077	4
9	15.298	11	17.226	8	3241.01	4	2343.911	7
10	18.019	13	17.964	4	2881.99	2	2213.550	2
11	14.594	11	16.258	7	3143.22	4	2162.850	6
12	17.603	10	17.202	7	3561.14	4	2259.450	4
13	18.233	11	20.752	3	2628.70	3	2065.724	3
14	18.767	13	12.196	5	3129.28	4	2247.266	4
15	16.676	8	17.484	2	3776.53	3	2265.176	3
16	14.404	4	21.910	7	2817.27	7	1979.851	2
17	21.809	8	18.715	5	4025.72	9	2227.952	3
18	14.456	12	18.402	3	3349.89	11	2176.611	2
19	15.804	10	17.938	6	3287.86	8	1954.326	2
20	19.494	11	18.922	5	2538.33	3	2271.901	6
21	16.936	10	14.203	4	4056.10	9	1946.981	4
22	17.473	7	14.519	3	2950.41	3	2212.840	3
23	10.698	4	16.074	8	4017.13	4	2046.632	4
24	15.104	6	18.651	7	3584.39	4	2266.475	4
25	18.760	10	19.017	8	3035.44	6	2172.965	3
26	19.572	4	14.376	6	3928.96	3	2235.301	5
27	16.517	10	21.805	3	2496.25	9	1962.889	2
28	20.253	12	15.160	6	3374.20	4	2333.199	7
29	22.102	7	19.081	2	4120.12	5	2500.466	3
30	15.462	12	13.978	3	3664.50	3	2630.580	7
31	17.111	9	15.177	5	3805.86	4	2189.025	3
Mean	9.42		5.13		4.74		**3.68**	
STD	2.66		1.98		2.42		**1.56**	

Table 4.36 Results of F2FWAC-II for wine dataset using Intra-cluster

Distance	Minimum				Maximum			
Metric	$\sqrt{n}$		*Stures*		$\sqrt{n}$		*Sturges*	
Fitness/number	Fit	K	Fit	K	Fit	K	Fit	K
1	19.00	6	18.32	3	238.01	11	4.81	3
2	16.80	12	18.92	5	309.91	9	4.71	3
3	16.15	7	18.56	3	474.23	13	4.88	4
4	14.83	12	19.14	7	251.76	4	4.82	7
5	17.71	12	21.56	5	261.83	12	4.94	3
6	17.08	8	14.73	4	294.45	10	4.66	6
7	16.14	13	18.29	7	319.59	11	4.82	8
8	15.86	12	20.45	7	560.95	6	4.79	3
9	19.66	9	18.47	6	215.72	10	4.53	7
10	17.36	12	17.84	3	200.29	10	4.70	5
11	16.94	9	16.07	7	228.61	11	4.88	4
12	16.57	12	17.44	7	521.95	7	4.68	2
13	13.80	11	24.65	8	308.23	10	4.80	6
14	16.84	10	18.72	7	516.90	12	4.81	8
15	17.03	10	16.06	4	403.53	11	4.68	3
16	16.07	5	16.19	3	301.38	11	4.51	8
17	18.50	7	15.09	3	253.43	4	4.86	2
18	12.70	3	17.83	8	345.05	5	4.74	7
19	18.01	11	20.27	7	398.89	10	4.54	5
20	15.88	6	17.53	4	212.57	13	5.49	5
21	15.22	11	20.66	6	293.21	11	5.18	4
22	17.98	5	17.95	8	243.34	9	4.79	7
23	19.44	11	22.04	4	241.23	13	4.81	7
24	16.91	4	19.16	4	231.19	11	4.86	2
25	17.52	9	22.72	6	413.77	11	4.77	5
26	16.52	8	17.89	7	295.44	10	4.81	2
27	16.59	9	17.89	7	268.80	7	4.81	7
28	15.22	11	14.68	4	449.19	9	4.86	3
29	20.91	6	18.18	8	301.70	4	4.63	6
30	15.29	12	16.17	7	494.05	11	5.43	6
31	18.18	12	16.50	3	278.35	3	4.52	4
Mean	9.19		5.55		9.32		**4.90**	
STD	2.82		1.82		2.88		**1.99**	

Table 4.37 Results of F2FWAC-II for wine dataset using Inter-cluster

Distance	Minimum				Maximum			
Metric	$\sqrt{n}$		*Sturges*		$\sqrt{n}$		*Sturges*	
Fitness/ number	Fit	K	Fit	K	Fit	K	Fit	K
1	15.103	2	13.040	2	3250.50	4	2498.819	2
2	16.502	12	16.259	3	2660.40	3	2026.681	3
3	12.970	8	15.835	3	2630.05	10	2210.111	4
4	18.355	11	23.634	4	2799.21	3	2623.008	8
5	24.142	3	16.849	7	2453.80	4	2118.064	4
6	14.695	10	22.084	7	2761.94	2	2211.811	4
7	19.898	9	16.567	6	2547.08	7	2374.102	4
8	18.463	10	17.805	8	3580.20	4	1764.140	3
9	16.557	12	17.907	4	3847.24	2	2366.682	4
10	12.981	5	18.394	7	3924.00	3	2021.334	5
11	16.069	12	14.214	2	3716.93	2	2270.198	4
12	15.678	13	18.082	6	3336.20	3	1491.888	2
13	17.763	12	17.748	5	3443.32	9	1652.390	2
14	21.166	5	21.635	8	2549.76	4	2150.413	5
15	19.291	10	21.836	7	2411.17	2	2318.678	8
16	16.546	6	17.239	5	3941.84	2	1749.983	2
17	15.881	7	18.018	8	3652.97	3	1798.990	6
18	12.422	2	18.976	5	2948.58	8	1890.532	3
19	16.903	12	14.923	6	3466.03	2	2437.527	2
20	17.168	12	18.003	2	2350.75	2	2199.763	7
21	14.584	10	18.360	8	3205.77	5	2474.275	3
22	18.208	7	22.824	8	2857.82	5	1910.584	3
23	15.100	4	14.932	2	2823.06	5	2533.212	4
24	15.759	11	20.271	7	2917.41	6	1734.281	3
25	19.195	10	18.205	3	2464.47	3	2348.078	3
26	15.084	10	16.703	7	3259.54	3	1726.920	2
27	18.277	9	21.173	4	2679.10	3	1349.153	2
28	17.925	10	14.546	2	2781.41	2	2431.423	5
29	15.845	12	22.051	6	3086.55	2	2046.924	3
30	20.169	9	16.994	8	3588.54	5	1428.320	2
31	18.890	4	17.879	5	2919.87	8	2213.994	4
Mean	8.68		5.32		4.06		**3.74**	
STD	3.29		2.15		2.26		**1.69**	

Table 4.38 Results of F2FWAC-III for wine dataset using Intra-cluster

Distance	Minimum				Maximum			
Metric	$\sqrt{n}$		*Sturges*		$\sqrt{n}$		*Sturges*	
Fitness/number	Fit	K	Fit	K	Fit	K	Fit	K
1	16.22	9	21.26	6	476.03	9	468.70	8
2	18.34	10	15.81	6	242.19	12	300.21	6
3	16.40	6	17.12	5	351.55	11	444.45	5
4	19.25	6	16.94	2	201.06	6	235.24	5
5	19.37	10	17.95	7	552.01	12	875.69	3
6	16.50	7	18.46	5	333.09	9	462.12	5
7	13.95	11	20.64	6	416.26	4	254.33	8
8	21.48	10	19.48	2	312.58	3	292.38	5
9	19.75	10	17.20	7	273.39	9	364.56	5
10	16.47	4	16.60	5	552.28	10	340.93	5
11	16.95	10	18.01	5	171.04	7	318.91	4
12	15.59	9	17.11	3	505.61	6	270.03	7
13	16.74	12	15.17	6	465.55	6	387.04	7
14	14.41	12	16.50	3	414.76	8	457.00	3
15	15.12	12	15.81	7	298.96	6	579.88	5
16	18.60	5	20.75	6	197.17	8	438.60	2
17	17.55	12	19.33	6	384.46	3	493.92	7
18	16.77	4	17.39	6	299.76	12	277.21	6
19	16.66	4	16.73	4	523.37	8	330.69	7
20	17.26	10	18.71	7	275.01	7	514.86	3
21	14.27	11	19.17	7	220.31	6	463.18	7
22	13.50	13	17.31	8	187.04	10	305.93	7
23	17.88	2	17.22	4	426.23	3	346.45	7
24	17.54	11	19.64	7	260.45	10	497.13	6
25	18.67	13	18.99	7	340.01	10	476.16	7
26	18.88	9	16.36	4	169.60	5	1104.06	7
27	15.69	11	16.86	6	417.17	2	897.47	8
28	15.46	10	21.21	2	330.81	5	478.99	5
29	17.27	7	19.40	6	296.91	11	446.56	8
30	16.93	11	15.35	2	335.40	5	285.47	5
31	13.89	13	16.44	7	556.21	4	483.96	7
Mean	9.16		**5.29**		7.32		5.81	
STD	3.03		**1.77**		2.96		1.64	

Table 4.39 Results of F2FWAC-III for wine dataset using Inter-cluster

Distance	Minimum				Maximum			
Metric	$\sqrt{n}$		*Sturges*		$\sqrt{n}$		*Sturges*	
Fitness/ number	Fit	K	Fit	K	Fit	K	Fit	K
1	15.864	11	16.681	5	3403.89	2	1978.422	3
2	14.726	5	17.156	6	3569.10	3	2595.135	2
3	17.698	11	16.651	7	3025.64	2	1919.507	2
4	17.092	9	19.679	7	3180.15	6	2003.342	4
5	16.923	5	21.328	5	3029.05	5	2422.256	3
6	15.676	13	19.259	7	3008.83	6	2503.906	2
7	18.767	12	20.299	6	3012.75	3	2273.016	2
8	18.157	7	18.902	6	3370.10	5	2115.659	2
9	19.050	8	19.807	3	3536.85	4	1573.959	3
10	15.154	8	22.696	7	3335.35	5	2311.131	6
11	16.946	2	11.930	7	3453.02	3	2036.198	2
12	15.546	6	19.622	4	3670.92	2	2652.912	6
13	16.690	9	16.019	7	2103.19	2	2100.335	3
14	14.109	8	16.474	3	2611.33	7	2094.132	2
15	13.411	12	21.048	8	3618.74	4	2279.808	4
16	16.612	13	18.233	8	3777.46	5	2372.357	3
17	17.961	11	12.107	2	3387.22	2	2051.100	5
18	17.142	5	10.783	4	2739.31	2	2447.559	2
19	20.210	12	19.183	4	2853.08	4	2103.669	2
20	20.841	9	24.361	2	3182.50	11	2576.203	4
21	19.592	10	18.843	6	3320.39	2	2324.492	5
22	20.576	6	12.529	2	3391.43	12	1939.277	3
23	13.609	2	18.931	3	3433.01	4	2094.433	3
24	17.712	11	16.470	6	3799.60	3	2606.266	8
25	18.559	3	18.190	3	4119.04	5	2389.364	3
26	15.958	7	15.260	3	3504.92	6	2350.802	3
27	14.621	6	18.048	8	3025.68	2	1878.096	3
28	9.383	2	18.475	5	4107.72	7	2168.266	3
29	21.460	8	16.163	3	3449.78	2	2107.143	5
30	17.960	13	18.724	8	3062.80	10	2030.347	3
31	16.269	13	16.589	5	4101.68	6	2198.080	8
Mean	8.29		5.16		4.58		**3.52**	
STD	3.48		1.98		2.69		**1.67**	

Table 4.40 Results of F2FWAC-I for WDBC dataset using Intra-cluster

Distance	Minimum				Maximum			
Metric	$\sqrt{n}$		Sturges		$\sqrt{n}$		Sturges	
Fitness/number	Fit	K	Fit	K	Fit	K	Fit	K
1	166.99	14	47.74	2	2064.8	15	2295.64	7
2	124.99	13	196.46	9	1103.5	9	2104.57	9
3	182.48	12	31.38	4	1847.4	19	1651.94	9
4	149.02	20	37.15	8	3234.6	15	1605.72	9
5	134.15	20	40.06	2	1806.9	15	3685.80	9
6	147.58	18	37.37	2	1066.5	14	1503.54	9
7	146.45	17	62.96	9	1249.3	20	2179.92	5
8	105.98	3	199.81	8	1747.3	22	1753.13	3
9	169.41	17	129.36	9	990.2	23	2581.49	10
10	130.87	15	31.99	3	1089.5	19	1747.94	3
11	121.36	13	197.73	5	1607.6	23	1593.81	5
12	142.96	3	23.81	10	1446.1	15	1434.09	8
13	164.95	18	28.29	10	1425.7	15	2431.98	8
14	140.19	20	27.91	5	1574.8	23	2282.61	7
15	151.17	18	161.31	10	949.6	18	1476.84	7
16	145.59	22	41.65	6	3988.0	22	1522.59	7
17	44.40	17	84.92	3	1781.8	20	3223.49	9
18	103.86	19	31.00	7	1399.0	14	1980.21	9
19	27.94	20	141.66	8	2112.5	11	1620.99	2
20	161.06	14	189.33	9	2509.0	8	1309.57	7
21	171.64	13	34.61	6	2128.5	3	1518.46	7
22	24.57	6	200.96	7	1805.4	5	1903.64	8
23	147.94	22	23.80	2	1700.7	3	1870.04	8
24	167.18	15	33.17	5	1429.9	7	2269.47	7
25	148.29	22	176.46	7	1114.8	23	1513.57	7
26	178.52	9	152.03	6	1842.5	24	1435.20	7
27	140.24	17	24.49	5	1722.4	2	1777.03	9
28	174.53	18	176.64	5	1274.1	12	1381.18	8
29	158.16	15	26.41	9	1644.4	17	4022.11	9
30	137.70	16	107.32	6	1255.8	14	2636.61	4
31	134.92	8	46.72	3	1056.7	19	2947.67	5
Mean	15.29		**6.11**		15.1		7.13	
STD	5.13		**2.51**		6.5		2.06	

Table 4.41 Results of F2FWAC-I for WDBC dataset using Inter-cluster

Distance	Minimum				Maximum			
Metric	$\sqrt{n}$		*Sturges*		$\sqrt{n}$		*Sturges*	
Fitness/number	Fit	K	Fit	K	Fit	K	Fit	K
1	47.420	11	22.729	9	24,789.99	10	14,429.959	3
2	166.361	21	29.225	4	32,395.38	10	14,490.710	6
3	32.166	7	155.201	10	31,642.81	22	11,441.616	3
4	144.455	9	26.803	4	17,865.52	7	12,555.768	3
5	162.154	12	30.984	6	26,703.41	12	12,161.380	3
6	31.202	15	24.101	4	28,504.65	9	12,449.388	2
7	129.127	21	40.151	5	22,991.42	5	11,385.558	4
8	43.900	2	161.001	9	30,095.13	7	7811.564	5
9	27.191	18	25.820	5	34,296.76	4	7576.805	3
10	134.438	23	154.693	9	24,052.10	8	12,867.891	3
11	145.383	21	43.186	3	22,772.45	5	12,186.043	2
12	22.540	3	157.522	8	30,734.72	6	11,091.987	3
13	27.641	9	45.981	6	25,403.83	16	11,709.717	3
14	185.525	3	36.748	3	29,413.33	22	12,026.947	4
15	30.404	18	173.074	6	32,652.73	7	10,392.370	5
16	167.429	16	164.509	10	28,662.81	3	13,103.299	3
17	39.593	10	34.025	6	27,756.19	16	11,594.310	3
18	137.342	15	23.628	6	25,005.70	11	14,504.883	4
19	146.083	21	22.073	8	27,252.12	7	14,004.863	7
20	139.191	23	201.558	9	25,357.25	5	12,031.828	3
21	160.851	15	23.475	4	22,935.84	4	11,769.932	2
22	106.104	22	35.019	5	16,906.11	3	10,418.284	5
23	149.586	18	25.957	4	26,477.87	5	11,323.763	2
24	143.424	16	28.741	3	21,923.83	4	11,866.223	2
25	153.057	20	28.152	4	21,438.96	2	10,634.021	4
26	139.925	19	20.690	9	18,231.91	4	9462.943	3
27	139.526	18	32.751	2	27,719.20	3	14,400.880	5
28	37.473	7	147.933	8	28,726.82	7	10,132.894	3
29	125.018	17	28.445	3	26,501.26	18	8003.931	3
30	149.412	20	191.117	9	28,576.04	3	11,036.472	4
31	47.241	2	152.141	9	28,702.52	6	10,256.760	2
Mean	14.58		6.13		8.10		**3.45**	
STD	6.58		2.49		5.49		**1.23**	

Table 4.42 Results of F2FWAC-II for WDBC dataset using Intra-cluster

Distance	Minimum				Maximum			
Metric	$\sqrt{n}$		*Sturges*		$\sqrt{n}$		*Sturges*	
Fitness/number	Fit	K	Fit	K	Fit	K	Fit	K
1	166.99	14	47.74	2	1366.53	18	1583.37	7
2	124.99	13	196.46	9	1526.30	12	1412.38	7
3	182.48	12	31.38	4	1741.72	22	1665.83	9
4	149.02	20	37.15	8	1078.81	7	2156.02	2
5	134.15	20	40.06	2	2051.59	15	1335.76	8
6	147.58	18	37.37	2	1302.25	21	1858.59	7
7	146.45	17	62.96	9	1157.75	20	2244.67	6
8	105.98	3	199.81	8	1631.65	23	2517.60	6
9	169.41	17	129.36	9	2502.03	17	3111.14	7
10	130.87	15	31.99	3	993.42	19	3707.71	5
11	121.36	13	197.73	5	1215.55	17	3871.41	10
12	142.96	3	23.81	10	1727.76	23	1618.43	7
13	164.95	18	28.29	10	2291.47	14	2743.12	3
14	140.19	20	27.91	5	1424.34	20	3530.70	10
15	151.17	18	161.31	10	2153.06	22	1411.43	9
16	145.59	22	41.65	6	1114.61	12	1577.33	7
17	44.40	17	84.92	3	2234.45	23	1732.44	8
18	103.86	19	31.00	7	1103.88	9	2462.71	9
19	27.94	20	141.66	8	1127.50	18	1432.78	5
20	161.06	14	189.33	9	1158.13	19	2199.71	8
21	171.64	13	34.61	6	1425.41	21	4190.21	4
22	24.57	6	200.96	7	1272.00	15	1771.61	10
23	147.94	22	23.80	2	1510.47	23	3178.54	10
24	167.18	15	33.17	5	1287.69	20	1555.71	7
25	148.29	22	176.46	7	2949.74	23	1746.53	7
26	178.52	9	152.03	6	934.61	13	2144.90	7
27	140.24	17	24.49	5	2459.39	21	2767.03	9
28	174.53	18	176.64	5	2323.62	21	2103.82	10
29	158.16	15	26.41	9	1339.02	20	2705.64	9
30	137.70	16	107.32	6	1103.42	13	2203.60	8
31	134.92	8	46.72	3	1107.63	11	1570.37	10
Mean	14.28		**6.13**		17.81		7.45	
STD	5.48		**2.63**		4.54		2.10	

Table 4.43 Results of F2FWAC-II for WDBC dataset using Inter-cluster

Distance	Minimum				Maximum			
Metric	$\sqrt{n}$		*Sturges*		$\sqrt{n}$		*Sturges*	
Fitness/ number	Fit	K	Fit	K	Fit	K	Fit	K
1	110.62	11	22.342	10	27,229.7	10	8466.58694	5
2	155.99	17	28.339	6	28,855.6	7	17,645.3055	3
3	31.21	15	27.102	7	29,180.4	3	9417.11741	2
4	168.68	16	170.740	7	27,330.9	20	7184.43	2
5	39.38	20	36.917	5	26,271.5	11	13,977.68	3
6	157.24	19	24.879	5	27,668.5	6	11,326.44	3
7	151.75	20	177.780	4	26,421.7	2	11,489.64	3
8	155.63	18	44.056	4	24,601.1	4	13,214.60	4
9	167.84	14	46.506	2	31,391.3	3	12,299.61	3
10	148.30	23	199.386	10	25,373.5	12	9341.53	2
11	154.00	13	25.696	5	25,014.1	8	12,382.66	3
12	157.95	10	35.139	5	33,228.7	2	10,011.10	7
13	142.73	14	183.216	4	25,970.6	3	11,726.12	6
14	44.04	4	23.596	7	32,984.2	8	12,281.69	4
15	146.16	22	172.321	7	28,953.1	22	11,978.28	2
16	75.02	2	25.984	8	29,665.5	23	12,112.20	7
17	144.31	13	153.383	9	29,503.8	2	13,589.23	8
18	37.99	9	55.946	2	26,636.5	2	14,133.22	6
19	125.33	18	40.643	3	29,032.4	10	8297.95	5
20	34.76	5	24.718	7	23,079.4	4	11,741.19	8
21	128.59	21	29.478	2	22,548.4	17	9232.28	3
22	141.30	18	25.350	6	30,881.2	9	11,800.67	2
23	27.12	6	29.761	3	26,197.8	3	11,087.35	3
24	30.22	3	103.974	2	29,947.6	20	10,930.48	5
25	137.92	23	23.517	2	21,213.6	2	11,272.47	2
26	33.87	5	49.521	3	26,437.3	21	10,427.14	5
27	27.97	3	36.306	4	25,724.5	4	10,898.12	3
28	145.67	19	31.455	6	29,172.8	5	9784.59	2
29	22.30	5	33.668	9	31,230.7	24	12,335.67	2
30	36.64	10	81.476	3	29,670.7	6	9563.90	3
31	27.46	7	135.345	9	30,848.9	17	12,030.31	3
Mean	13.01		5.35		9.35		**3.84**	
STD	6.68		5.35		7.35		**1.86**	

Table 4.44 Results of F2FWAC-III for WDBC dataset using Intra-cluster

Distance	Minimum				Maximum			
Metric	$\sqrt{n}$		*Sturges*		$\sqrt{n}$		*Sturges*	
Fitness/number	Fit	K	Fit	K	Fit	K	Fit	K
1	135.1	16	165.22	7	1050.88	23	1949.01	9
2	22.7	15	164.51	8	1807.10	9	2729.97	9
3	142.2	18	19.77	7	1029.39	17	1693.59	10
4	31.3	13	28.15	3	1612.35	16	2411.96	7
5	151.5	17	41.13	3	2290.10	8	1842.71	6
6	28.0	6	27.14	3	3658.85	12	1868.43	8
7	142.7	12	38.07	4	902.43	11	1681.79	6
8	135.6	18	26.27	5	1101.52	16	1539.75	2
9	128.2	24	24.21	9	1591.89	18	2409.39	3
10	30.3	6	19.95	7	1074.86	16	2597.87	7
11	27.9	3	36.03	6	1290.74	13	1625.82	8
12	46.7	2	26.18	5	1250.54	16	2038.73	10
13	129.4	23	146.19	7	3541.85	15	1522.65	8
14	161.4	14	33.43	5	2068.93	16	3966.94	2
15	35.7	5	172.05	9	1103.69	21	1237.60	10
16	142.3	22	46.10	4	938.46	22	2045.48	7
17	166.6	24	25.32	2	1177.20	23	2163.63	7
18	28.3	3	129.36	6	1814.70	22	3779.64	3
19	123.0	22	47.62	6	1378.34	15	1713.49	2
20	41.4	3	29.28	3	1595.04	15	1231.13	9
21	43.3	11	45.34	3	1277.26	24	2144.29	8
22	39.4	12	194.78	6	1293.09	24	3239.07	5
23	167.0	4	39.96	10	2230.66	8	3162.83	5
24	26.2	10	177.05	7	2314.41	10	1949.79	8
25	154.6	24	42.16	4	1259.97	23	1890.04	9
26	138.6	21	24.62	7	1073.02	16	2766.33	5
27	138.0	22	202.51	8	1705.94	16	1977.65	9
28	149.0	13	28.83	5	1607.35	18	1295.03	3
29	152.8	4	27.37	7	1309.23	16	1743.36	6
30	37.5	12	29.70	5	1879.10	15	2351.17	9
31	130.9	24	28.26	2	1177.11	15	3399.80	5
Mean	13.6		5.58		16.42		6.61	
STD	7.6		2.14		4.63		2.54	

Table 4.45 Results of F2FWAC-III for WDBC dataset using Inter-cluster

Distance	Minimum				Maximum			
Metric	$\sqrt{n}$		*Sturges*		$\sqrt{n}$		*Sturges*	
Fitness/ number	Fit	K	Fit	K	Fit	K	Fit	K
1	47.420	11	48.588	5	27,071.11	7	9114.684	4
2	166.361	21	27.301	3	16,460.23	3	13,117.992	4
3	32.166	7	203.507	7	28,975.14	4	10,411.008	5
4	144.455	9	176.595	10	30,320.72	2	11,162.007	2
5	162.154	12	25.973	3	26,071.14	4	7217.242	3
6	31.202	15	25.033	4	27,488.38	5	12,692.544	8
7	129.127	21	188.110	8	28,239.24	6	13,193.462	4
8	43.900	2	162.981	5	32,456.98	24	10,694.813	2
9	27.191	18	37.349	4	24,245.43	5	8373.473	3
10	134.438	23	32.271	2	28,862.22	8	13,531.131	6
11	145.383	21	41.329	4	28,628.58	5	11,833.871	2
12	22.540	3	23.480	6	28,125.31	8	11,409.729	2
13	27.641	9	33.370	4	27,172.91	15	7917.673	2
14	185.525	3	31.553	6	28,260.13	8	13,487.203	9
15	30.404	18	171.719	8	25,354.71	18	11,026.349	4
16	167.429	16	27.022	8	27,827.28	2	9956.647	8
17	39.593	10	173.416	7	24,028.88	2	7532.554	5
18	137.342	15	40.001	7	26,494.20	18	13,769.499	5
19	146.083	21	83.894	3	25,946.40	5	7490.365	3
20	139.191	23	159.113	10	28,372.86	24	11,117.693	2
21	160.851	15	40.991	4	26,948.81	18	11,690.336	3
22	106.104	22	163.540	9	27,961.21	10	9821.677	7
23	149.586	18	26.797	4	26,525.58	12	12,523.558	3
24	143.424	16	43.846	4	33,508.60	24	11,181.053	2
25	153.057	20	191.573	6	21,949.66	2	9699.145	2
26	139.925	19	23.205	6	25,707.94	8	14,673.479	2
27	139.526	18	31.894	2	26,646.72	4	7791.942	3
28	37.473	7	138.424	3	26,183.57	3	12,928.639	2
29	125.018	17	30.407	5	31,652.23	8	10,391.139	4
30	149.412	20	34.273	8	27,158.27	7	12,978.711	8
31	47.241	2	25.257	8	21,428.04	6	14,193.305	10
Mean	14.58		5.58		8.87		4.16	
STD	6.58		2.29		6.83		2.38	

Table 4.46 Results of the classification percentage with T1FIS for Iris dataset

FIS type	Mamdani			Sugeno		
Membership function	Triangular	Gaussian	Trapezoidal	Triangular	Gaussian	Trapezoidal
1	73.33	83.33	66.67	80	75.86	70
2	70	80	66.67	76.67	89.66	70
3	83.33	86.67	76.67	73.33	86.67	63.33
4	80	90	73.33	73.33	86.67	66.67
5	76.67	73.33	76.67	73.33	79.31	80
6	70	83.33	66.67	73.33	89.66	73.33
7	76.67	83.33	66.67	76.67	89.66	63.33
8	76.67	86.67	76.67	76.67	75.86	66.67
9	73.33	76.67	73.33	63.33	86.21	66.67
10	73.33	86.67	60	73.33	75.86	63.33
11	70	76.67	73.33	76.67	86.67	66.67
12	73.33	86.67	70	63.33	86.67	66.67
13	73.33	83.33	73.33	70	86.21	73.33
14	76.67	80	73.33	63.33	86.67	70
15	76.67	73.33	70	76.67	86.67	63.33
16	73.33	83.33	76.67	70	89.66	70
17	80	73.33	66.67	80	86.67	66.67
18	73.33	80	73.33	76.67	86.21	66.67
19	70	80	70	70	79.31	66.67
20	73.33	86.67	76.67	70	79.31	73.33
21	76.67	83.33	70	70	86.67	63.33
22	80	86.67	73.33	80	89.66	66.67
23	83.33	80	73.33	80	86.67	73.33
24	76.67	86.67	66.67	76.67	89.66	63.33
25	86.67	83.33	63.33	70	89.66	73.33
26	73.33	76.67	66.67	70	75.86	66.67
27	76.67	76.67	66.67	76.67	75.86	60
28	76.67	86.67	76.67	70	89.66	73.33
29	73.33	70	73.33	80	86.21	70
30	76.67	83.33	66.67	70	79.31	63.33
31	70	73.33	76.67	73.33	75.86	73.33
Mean	75.59	**81.29**	70.97	73.33	**84.34**	68.17
STD	4.16	**5.21**	4.57	4.87	**5.23**	4.46

Table 4.47   Results of the classification percentage with T1FIS for wine dataset

FIS type	Mamdani			Sugeno		
Membership function	Triangular	Gaussian	Trapezoidal	Triangular	Gaussian	Trapezoidal
1	60	63.89	44.44	33.33	71.43	44.44
2	38.89	69.44	44.44	80	52.78	66.67
3	60	65.71	66.67	38.89	55.56	44.44
4	72.22	68.57	66.67	69.44	52.78	66.67
5	33.33	69.44	44.44	38.89	72.22	44.44
6	38.89	72.22	44.44	36.11	72.22	66.67
7	60	66.67	66.67	69.44	55.56	65.71
8	38.89	52.78	44.44	38.89	72.22	44.44
9	38.89	66.67	44.44	38.89	72.22	44.44
10	71.43	47.22	44.44	69.44	72.22	65.71
11	60	50	44.44	60	62.86	66.67
12	72.22	55.56	65.71	38.89	69.44	44.44
13	38.89	74.29	44.44	71.43	60	47.22
14	36.11	69.44	44.44	38.89	62.86	44.44
15	71.43	65.71	44.44	69.44	62.86	44.44
16	69.44	66.67	65.71	38.89	72.22	44.44
17	38.89	65.71	44.44	38.89	55.56	44.44
18	72.22	50	68.57	38.89	50	66.67
19	38.89	69.44	47.22	38.89	69.44	44.44
20	38.89	66.67	44.44	69.44	72.22	44.44
21	40	65.71	44.44	38.89	72.22	44.44
22	38.89	61.11	65.71	61.11	72.22	44.44
23	38.89	66.67	66.67	38.89	75	44.44
24	36.11	68.57	44.44	38.89	72.22	47.22
25	40	72.22	45.71	60	69.44	44.44
26	36.11	66.67	44.44	33.33	72.22	44.44
27	38.89	66.67	45.71	38.89	69.44	66.67
28	38.89	36.11	65.71	33.33	52.78	44.44
29	69.44	65.71	45.71	69.44	72.22	66.67
30	38.89	44.44	44.44	36.11	71.43	66.67
31	38.89	69.44	66.67	72.22	74.29	66.67
Mean	48.53	**63.21**	51.76	49.62	**66.46**	52.45
STD	14.46	**9.16**	10.35	15.69	**7.93**	10.61

Table 4.48 Results of the classification percentage with T1FIS for WDBC dataset

FIS type	Mamdani			Sugeno		
Membership function	Triangular	Gaussian	Trapezoidal	Triangular	Gaussian	Trapezoidal
1	71.93	92.98	71.93	71.05	92.11	93.86
2	71.93	92.98	71.05	71.93	92.04	93.86
3	71.93	92.98	71.93	71.93	92.11	93.86
4	71.93	90.35	71.05	72.81	92.11	93.81
5	71.93	89.47	71.93	71.93	92.04	93.86
6	71.93	93.86	71.05	91.23	92.11	93.86
7	71.93	91.23	71.93	72.57	92.11	93.86
8	71.93	91.23	72.81	71.93	92.11	93.86
9	71.05	92.11	91.23	71.93	92.11	93.86
10	71.93	92.11	71.93	71.93	92.11	93.86
11	71.05	93.86	71.05	71.93	93.86	93.86
12	71.93	92.11	71.93	71.93	92.11	93.86
13	71.93	92.11	71.05	71.05	93.86	93.86
14	71.93	93.86	71.93	72.81	92.11	93.86
15	71.93	92.98	91.23	71.93	93.86	93.86
16	71.93	92.98	71.93	71.93	94.74	93.86
17	71.93	92.11	71.93	72.81	92.11	93.86
18	71.93	94.74	91.23	72.81	93.86	93.86
19	71.93	92.11	71.93	71.93	92.11	93.86
20	71.93	91.23	72.57	71.93	92.98	93.86
21	71.93	92.98	71.93	71.05	92.11	93.86
22	71.93	92.98	71.93	71.93	92.11	93.86
23	71.93	92.11	71.93	71.93	92.11	93.86
24	72.81	89.47	72.81	71.93	92.11	93.86
25	71.93	92.11	71.93	71.05	92.11	93.86
26	71.93	92.11	71.93	71.93	92.11	93.86
27	71.05	91.23	71.05	71.93	92.11	93.86
28	71.93	93.86	72.81	71.05	92.11	93.86
29	71.05	92.98	73.68	91.23	92.11	93.86
30	71.93	92.98	71.93	71.93	92.11	95.61
31	71.93	92.11	71.93	71.93	92.11	93.86
Mean	71.84	**92.33**	73.79	73.17	92.44	**93.91**
STD	0.35	**1.22**	5.83	4.85	0.74	**0.32**

4.2.2 *Interval Type 2 Fuzzy Logic Classifiers (IT2FIS)*

The best option for the IT2FIS for the Iris dataset is to use Trapezoidal MFs. The results of the IT2FIS show a mean of 71.61% and 76.02% with Mamdani and Sugeno type, respectively; in Table 4.49 the results are shown. Table 4.50 show the results of the IT2FIS for the wine dataset; the best result with a mean of 72.34% for the Mamdani type using Triangular MFs, this value is marked in bold type.

An IT2FIS of Sugeno type with Gaussian MFs is the best result obtained for the WDBC dataset. The mean classification of IT2FIS Sugeno is 94.65%. Table 4.51 shows the classification with IT2FIS for WDBC.

4.3 Statistical Comparison Results of the Optimization of the Fuzzy Classification Models

We also perform a statistical comparison of all the results obtained the proposed model for the section of classification methods. The statistical test used for comparison is the z-test, whose parameters are defined in Table 4.52. In applying the statistic z-test, with significance level of 0.05, and the alternative hypothesis stating that the μ_1 is greater than the μ_2 ($\mu_1 > \mu_2$), and of course the null hypothesis tells us that the μ_1 is lower than or equal to the μ_2 ($\mu_1 \leq \mu_2$), with a rejection region for all values that fall above 1.645. We are presenting 31 experiments with the same parameters and conditions for the T1FIS and IT2FIS for this work, so the n_1 and n_2 are equal 31.

The main objective of applying the statistical z-test is to analyze the performance and thus find if there is significant evidence of the proposed model results with Interval Type 2 Fuzzy Logic (IT2FIS) being better of the Type 1 Fuzzy Inference Systems (T1FIS). The classification percentage generated for each FIS. The results of the statistical z-tests are shown from Tables 4.53, 4.54, 4.55, 4.56, 4.57 and 4.58, so there is significant evidence to reject the null hypothesis because the value of $p < 0.05$ and the value of $z > 1.645$ and we accepted the alternative hypothesis. Therefore, the results obtained of the the classification percentage of Interval Type 2 Fuzzy Inference System are better than the Type 1 Fuzzy Inference System.

Table 4.53 shows the Z-test of the Mamdani type FISs for the Iris dataset. In the three z-tests the evidence is not significant. For the wine dataset, 2 of 3 Z-tests shows the significant evidence; the results are shown in Table 4.54.

Table 4.55 presents 2 of 3 Z-tests for the WDBC dataset with significant evidence of the Mandani type FISs. As the same way, Table 4.56 shows 2 of 3 Z-tests with significant evidence of the Sugeno type FISs but for the Iris dataset.

Z-tests results indicate that 2 of 3 with significant evidence for the wine dataset of the Sugeno type FISs, in Table 4.57 the results are shown.

For the WDBC dataset of the Sugeno type FISs, the Z-tests yielded only 1 of 3 with significant evidence (see Table 4.58).

Table 4.49 Results of the classification percentage with IT2FIS for Iris dataset

IT2FIS type	Mamdani			Sugeno		
Membership function	Triangular	Gaussian	Trapezoidal	Triangular	Gaussian	Trapezoidal
1	83.33	76.67	70	70	55.17	50
2	80	56.67	70	66.67	90	86.67
3	76.67	53.33	63.33	70	86.21	80
4	56.67	56.67	66.67	76.67	55.17	60
5	76.67	70	70	70	37.93	80
6	53.33	70	66.67	70	34.48	50
7	73.33	66.67	70	70	34.48	66.67
8	53.33	53.33	40	70	36.67	56.67
9	83.33	76.67	70	80	37.93	56.67
10	83.33	70	40	83.33	75.86	60
11	53.33	53.33	66.67	80	56.67	60
12	83.33	73.33	66.67	80	58.62	80
13	83.33	50	70	70	46.67	60
14	86.67	53.33	63.33	70	75.86	66.67
15	73.33	53.33	66.67	70	46.67	70
16	76.67	70	70	83.33	93.1	60
17	73.33	50	70	70	36.67	80
18	83.33	66.67	70	76.67	36.67	56.67
19	76.67	70	53.33	70	33.33	56.67
20	53.33	53.33	70	86.67	31.03	53.33
21	83.33	70	70	83.33	55.17	63.33
22	53.33	50	70	66.67	34.48	56.67
23	86.67	66.67	40	83.33	83.33	63.33
24	43.33	53.33	40	86.67	83.33	53.33
25	76.67	53.33	40	86.67	55.17	60
26	90	70	40	76.67	36.67	46.67
27	43.33	66.67	70	80	37.93	66.67
28	76.67	56.67	70	70	56.67	56.67
29	46.67	53.33	63.33	80	41.38	63.33
30	73.33	60	40	80	75.86	60
31	83.33	53.33	70	80	56.67	50
Mean	**71.61**	61.18	61.51	**76.02**	54.06	62.26
STD	**14.42**	8.84	12.29	**6.52**	19.50	10.09

Table 4.50 Results of the classification percentage with IT2FIS for wine dataset

IT2FIS type	Mamdani			Sugeno		
Membership function	Triangular	Gaussian	Trapezoidal	Triangular	Gaussian	Trapezoidal
1	72.22	38.89	66.67	63.89	69.44	63.89
2	72.22	61.11	69.44	63.89	69.44	66.67
3	72.22	36.11	69.44	58.33	69.44	63.89
4	72.22	36.11	66.67	61.11	68.57	72.22
5	72.22	63.89	69.44	63.89	69.44	61.11
6	72.22	33.33	60	72.22	68.57	69.44
7	72.22	28.57	69.44	60	69.44	63.89
8	72.22	60	60	58.33	69.44	63.89
9	72.22	60	69.44	63.89	68.57	72.22
10	72.22	61.11	69.44	55.56	69.44	63.89
11	72.22	45.71	72.22	61.11	69.44	58.33
12	72.22	57.14	69.44	61.11	69.44	69.44
13	72.22	30.56	68.57	51.43	68.57	63.89
14	68.57	38.89	69.44	61.11	72.22	69.44
15	74.29	60	69.44	62.86	69.44	63.89
16	72.22	57.14	69.44	61.11	72.22	63.89
17	72.22	38.89	68.57	60	69.44	63.89
18	72.22	63.89	72.22	63.89	69.44	69.44
19	72.22	77.78	69.44	63.89	69.44	69.44
20	72.22	63.89	66.67	72.22	63.89	66.67
21	74.29	36.11	69.44	55.56	69.44	69.44
22	74.29	38.89	66.67	58.33	69.44	62.86
23	72.22	33.33	69.44	61.11	68.57	58.33
24	77.14	63.89	69.44	61.11	69.44	69.44
25	72.22	30.56	69.44	61.11	72.22	66.67
26	72.22	38.89	69.44	58.33	63.89	58.33
27	72.22	63.89	69.44	72.22	69.44	69.44
28	72.22	36.11	66.67	72.22	72.22	75
29	72.22	45.71	60	61.11	68.57	69.44
30	72.22	61.11	69.44	61.11	68.57	58.33
31	68.57	45.71	69.44	66.67	69.44	69.44
Mean	**72.34**	48.62	68.20	62.22	**69.24**	66.00
STD	**1.45**	13.74	3.03	4.92	**1.79**	4.42

Table 4.51 Results of the classification percentage with IT2FIS for WDBC dataset

IT2FIS type	Mamdani			Sugeno		
Membership function	Triangular	Gaussian	Trapezoidal	Triangular	Gaussian	Trapezoidal
1	85.965	80.7	85.96	64.91	93.81	58.77
2	85.088	89.47	89.47	62.28	94.69	56.14
3	85.088	85.96	86.84	81.42	94.74	53.51
4	85.088	85.96	87.72	64.91	93.86	56.14
5	85.965	91.23	85.96	62.28	94.69	58.77
6	85.088	86.84	89.47	81.42	94.74	55.26
7	85.965	88.6	85.96	67.54	94.74	59.65
8	85.965	86.84	85.08	61.4	94.74	53.51
9	85.088	85.96	85.08	59.65	94.74	53.1
10	85.088	86.84	89.47	57.89	94.74	51.75
11	85.965	85.96	89.47	61.4	94.74	51.75
12	85.965	89.47	89.47	58.77	93.86	61.4
13	85.088	89.47	85.96	63.16	94.74	54.39
14	85.088	85.96	85.08	60.53	94.74	54.39
15	85.088	89.47	85.96	61.4	94.69	57.02
16	85.965	85.96	89.47	64.91	94.74	61.4
17	85.965	85.96	86.84	59.65	94.74	54.39
18	85.088	85.96	87.72	58.77	94.74	57.89
19	85.088	85.96	85.96	62.83	94.74	50.88
20	84.956	87.61	89.47	58.77	94.74	58.77
21	85.088	85.96	85.96	58.77	94.74	53.51
22	85.965	85.96	89.47	61.4	94.74	52.63
23	85.088	89.47	86.84	58.77	94.74	61.4
24	85.088	86.84	87.72	62.28	94.74	53.51
25	85.965	87.72	89.47	60.53	94.74	53.51
26	85.088	87.72	89.47	60.18	94.74	57.02
27	85.088	85.96	85.96	64.04	94.69	57.02
28	85.841	92.04	89.47	64.91	94.74	77.19
29	85.088	87.72	86.84	64.04	94.74	53.51
30	85.088	87.72	89.47	61.4	94.74	55.26
31	85.088	85.96	89.47	59.65	94.74	58.77
Mean	85.39	87.20	**87.63**	62.90	**94.65**	56.52
STD	0.43	2.09	**1.72**	5.48	**0.27**	4.85

Table 4.52 Parameters of Z-test

Parameter	Value
Confidence interval	95%
Significance level (α)	5%
Null hypothesis (H_0)	$^*\mu_1 \leq \mu_2^*$
Alternative hypothesis (H_a)	$\mu_1 > \mu_2$
Critical value	1.645

μ_1 = Average the classification percentage of the Interval Type 2 Fuzzy Inference System
μ_2 = Average the classification percentage of the Type 1 Fuzzy Inference System

Table 4.53 Z-test of Mamdani type for Iris dataset

IT2FIS		T1FIS		Parameters		Evidence
μ_1	σ_1	μ_2	σ_2	z	$p < 0.05$	
Fuzzy classifiers using triangular MFs						
71.61	14.42	75.59	4.16	1.47	0.9301	Not significant
Fuzzy classifiers using Gaussian MFs						
61.18	8.84	81.29	5.21	− 10.91	1	Not significant
Fuzzy classifiers using trapezoidal MFs						
61.51	12.29	70.97	4.57	− 4.017	1	Not significant

Table 4.54 Z-test of Mamdani type for wine dataset

IT2FIS		T1FIS		Parameters		Evidence
μ_1	σ_1	μ_2	σ_2	z	$p < 0.05$	
Fuzzy classifiers using triangular MFs						
72.34	1.45	48.53	14.46	9.12	0	Significant
Fuzzy classifiers using Gaussian MFs						
48.62	13.74	63.21	9.16	− 5.526	1	Not significant
Fuzzy classifiers using trapezoidal MFs						
68.2	3.03	51.76	10.35	8.48	0	Significant

Table 4.55 Z-test of Mamdani type for WDBC dataset

IT2FIS		T1FIS		Parameters		Evidence
μ_1	σ_1	μ_2	σ_2	z	$p < 0.05$	
Fuzzy classifiers using triangular MFs						
85.5	0.43	71.84	0.35	137.17	0	Significant
Fuzzy classifiers using Gaussian MFs						
87.2	2.09	92.33	1.22	− 11.80	1	Not significant
Fuzzy classifiers using trapezoidal MFs						
87.63	1.72	73.79	5.83	12.67	0	Significant

Table 4.56 Z-test of Sugeno type for Iris dataset

IT2FIS		T1FIS		Parameters		Evidence
μ_1	σ_1	μ_2	σ_2	z	$p < 0.05$	
Fuzzy classifiers using triangular MFs						
76.02	6.52	73.33	4.87	1.84	0.0329	Significant
Fuzzy classifiers using Gaussian MFs						
54.06	19.5	84.34	5.23	− 8.351	1	Not significant
Fuzzy classifiers using trapezoidal MFs						
62.26	10.09	68.17	4.46	− 2.98	0.9986	Not significant

Table 4.57 Z-test of Sugeno type for wine dataset

IT2FIS		T1FIS		Parameters		Evidence
μ_1	σ_1	μ_2	σ_2	z	$p < 0.05$	
Fuzzy classifiers using triangular MFs						
62.22	4.92	49.62	15.69	4.26	0	Significant
Fuzzy classifiers using Gaussian MFs						
69.24	1.79	66.46	7.93	1.90	0.0285	Significant
Fuzzy classifiers using trapezoidal MFs						
66	4.42	52.45	10.61	6.56	0	Significant

Table 4.58 Z-test of Sugeno type for WDBC dataset

IT2FIS		T1FIS		Parameters		Evidence
μ_1	σ_1	μ_2	σ_2	z	$p < 0.05$	
Fuzzy classifiers using triangular MFs						
62.9	5.48	73.17	4.85	− 7.81	1	Not Significant
Fuzzy classifiers using Gaussian MFs						
94.65	0.27	92.44	0.74	15.62	0	Significant
Fuzzy classifiers using trapezoidal MFs						
56.52	4.85	93.91	0.32	− 42.8	1	Not significant

Based on the statistical z-test results, we can make the conclusion that the results obtained of the optimization of fuzzy integrators with the IT2FIS are better than the T1FIS for some datasets. IT2FIS of Mamdani type is better than T1FIS of Mamdani type for Wine and WDBC datasets, on the contrary for the Iris dataset; the hypothesis tests do not offer significant evidence. The IT2FIS of Sugeno type is better than T1FIS of Sugeno type for the wine dataset, but, for Iris and WDBC datasets is only better with Triangular and Gaussian Membership Functions, respectively.

Chapter 5
Conclusions to the Hybrid Method Between Fireworks Algorithm and Competitive Neural Network

In this book, the proposed method is sectioned in two parts to achieve the main goal of design a hybrid method of competitive learning. In the first section, FWA adaptation (FWAC, FFWAC and F2FWAC) as a method for optimizing the optimal number of clusters based on the number of centroids for clustering problems is presented; and, Type 1 (T1FIS) and Interval Type 2 Fuzzy Inference Systems (IT2FIS) joined a Competitive Neural Network for classification problems, in the second section.

For the first section (Optimization methods), according to the results, the optimizations of the number of clusters with FWAC are very variable; although in some cases the optimizations are satisfactory. For example, when the evaluations of FWAC are with Inter-cluster validation and the Sturges Law, we achieve a good approximation for the Iris and Wine datasets with minimum and maximum distance.

With the Fuzzy Fireworks Algorithm for Clustering, the results of FFWAC-I, II and III are more consistent, in the majority of the cases when we used Inter-cluster as a validation measure with Sturges Law, in the data set of Iris and Wine the approximation the number of clusters is satisfactory; moreover, when the number of clusters is the Square root of N, the optimizations are satisfactory in few cases, as example, we can find the results of FFWA-II with Gaussians Membership Functions and FFWA-III with Trapezoidal Membership Functions.

The results obtained with Interval Type 2 Fuzzy Fireworks Algorithm for Clustering show an unanimous decision for the combination of the metrics for the three different datasets tested, i.e., the combination is using Sturges Law with Inter-cluster and maximum distance.

In the comparison among optimization methods, the best results of the mean for the IRIS dataset is using the combination of the maximum distance with Inter-cluster and Sturges Law in FFWAC-III; in the WINE dataset, the combination is to use the minimum distance with Inter-cluster and Sturges Law in FWAC; and finally, for the WDBC, the combination the minimum distance with Inter-cluster and Sturges Law in FWAC-I is used.

© The Author(s), under exclusive license to Springer Nature Switzerland AG 2023
F. Valdez et al., *Hybrid Competitive Learning Method Using the Fireworks Algorithm and Artificial Neural Networks*, SpringerBriefs in Computational Intelligence,
https://doi.org/10.1007/978-3-031-47712-6_5

For the second section of proposed method, the results obtained are classified into three parts:

1. For the Iris dataset, the best result obtained from the Fuzzy Inference System is using Gaussian Membership Functions in the variables into a T1FIS of Mamdani type, with 90% of correct classification. But, the best mean of the variations is using the Gaussian Membership Functions into a T1FIS of Sugeno type with 84.34%.
2. For the Wine dataset, the best results of all variations was with an IT2FIS of the Mamdani type using Gaussian Membership Functions with 72.34%, but, the best FIS obtained was using Triangular Membership Functions into a T1FIS of the Sugeno type, with 80% of correct classification.
3. The 95.61% of correct classification is obtained for the best designed T1FIS of Sugeno type, and, the best mean is 94.65%; was obtained with an IT2FIS of Sugeno type with Trapezoidal Membership Functions for the WDBC dataset.

A general conclusion for the use of proposed hybrid method is to use the following parameters/metrics: Sturges Law with as a metric of upper bound, Inter-cluster with maximum distance as cluster validation metric and, Gaussian MFs into the FIS variables of the FFWAC, this, for the first section named as Optimization methods. For the second section, named as Classification methods, the best option is to use Interval Type 2 Fuzzy Inference Systems of Sugeno type.

The objectives and goals in this book were achieved satisfactorily, because the results reported from experiments are efficient; a good optimal number of centroids in the first section, and, good correct classifications percentages in the second section.

5.1 Future Works

In future work, it could be interesting to apply the proposed hybrid method in other data sets, as well as in other application areas such as control systems, pattern recognition and to obtain prediction of time series data. In addition, to improve the methods for clustering problems; it could include the implementation of different optimization algorithms and other types of artificial neural networks.

We also could try to design generalized type 2 Fuzzy Systems to apply in the same areas mentioned above, or, in another complex problems.

Index

© The Author(s), under exclusive license to Springer Nature Switzerland AG 2023   103
F. Valdez et al., *Hybrid Competitive Learning Method Using the Fireworks Algorithm and Artificial Neural Networks*, SpringerBriefs in Computational Intelligence,
https://doi.org/10.1007/978-3-031-47712-6

Printed in the United States
by Baker & Taylor Publisher Services